Micropropagation &

Rukhama Haq
Khadija Gilani
Shagufta Naz

Micropropagation & Callogenesis of Vinca rosea. L.

comparison of in vivo and in vitro response of antimicrobial activity of Sadabahar

LAP LAMBERT Academic Publishing

Impressum/Imprint (nur für Deutschland/only for Germany)
Bibliografische Information der Deutschen Nationalbibliothek: Die Deutsche Nationalbibliothek verzeichnet diese Publikation in der Deutschen Nationalbibliografie; detaillierte bibliografische Daten sind im Internet über http://dnb.d-nb.de abrufbar.

Coverbild: www.ingimage.com

Verlag: LAP LAMBERT Academic Publishing GmbH & Co. KG
Heinrich-Böcking-Str. 6-8, 66121 Saarbrücken, Deutschland
Telefon +49 681 3720-310, Telefax +49 681 3720-3109
Email: info@lap-publishing.com

Approved by: Lahore, Lahore College for Women University, August.,2009

Herstellung in Deutschland:
Schaltungsdienst Lange o.H.G., Berlin
Books on Demand GmbH, Norderstedt
Reha GmbH, Saarbrücken
Amazon Distribution GmbH, Leipzig
ISBN: 978-3-8484-3817-4

Imprint (only for USA, GB)
Bibliographic information published by the Deutsche Nationalbibliothek: The Deutsche Nationalbibliothek lists this publication in the Deutsche Nationalbibliografie; detailed bibliographic data are available in the Internet at http://dnb.d-nb.de.

Cover image: www.ingimage.com

Publisher: LAP LAMBERT Academic Publishing GmbH & Co. KG
Heinrich-Böcking-Str. 6-8, 66121 Saarbrücken, Germany
Phone +49 681 3720-310, Fax +49 681 3720-3109
Email: info@lap-publishing.com

Printed in the U.S.A.
Printed in the U.K. by (see last page)
ISBN: 978-3-8484-3817-4

Comparison of *in vitro* response of micropropagation, callogenesis and antimicrobial activity of *Vinca rosea*. L.

Dedicated to my Parents

TABLE OF CONTENTS

List of Tables

List of Plates

List of Graphs

List of Abbreviations

2, 4-D	2, 4- dichlorophenoxyacetic acid
BAP	6- benzyl amino purine
NAA	Naphthalene acetic acid
&	And
cm	Centimetre
~	Approximately
°C	Degree Celsius
DMRT	Duncan's Multiple Range Test
EDTA	Ethylene diamine tetra acetate
g/l	gram/litre
hrs	Hours
HCl	Hydrochloric acid
IAA	Indole 3- acetic acid
IBA	Indole 3 butyric acid
Kin	Kinetin
LCWU	Lahore college for women university
LAFC	Laminar air flow cabinet
LSD	Least significant difference
Ltd.	Limited
l	Litre
μ	Micro
μl	Microlitre
mg	Milligram
mg/ml	Milligram per millilitre
ml	Millilitre
mm	Millimetre
MIC	Minimum inhibitory concentration
min	Minutes
M	Molar
MS	Murashige and Skoog (1962)
N	Normal solution
No.	Number
%	Percentage
KOH	Potassium hydroxide
lb/inch^2	Pounds per square inch
pH	Power of Hydrogen ion concentration
ppm	Plant preservative mixture
Pvt.	Private
<p>	Probability
rpm	Revolution per minute
Sr.	Serial
±SD	Standard deviation among triplicates
X	Strength
UV	Ultraviolet
v/v	Volume per volume
w/v	Weight per volume
WHO	World Health Organization

ABSTRACT

Vinca rosea L. Syn. *Catharanthus roseus* L. G. Don. is an important medicinal plant which belongs to the family Apocynaceae. The present study deals with a simple and efficient protocol for *in vitro* micropropagation of *Vinca rosea* through shoot tip and nodal portion explants and callogenesis of *Vinca rosea* through leaf, node and fruit explants. Explants were cultured on MS basal medium and supplemented with different concentrations of BAP (mg/l) and NAA (mg/l) for micropropagation whereas 2, 4-D (mg/l) and Kin (mg/l) for callogenesis. Multiple shoots were obtained on all the concentrations of BAP and NAA, but BAP (1 mg/l) showed best response (90% & 80%) from both the explants. Similarly best callus response was observed on MS medium supplemented with 2, 4- D + Kin (1+1 mg/l) in all the explants (95% from leaf, 80% from node & 60% from fruit). A protocol for the cell suspension culture have also been established in the present study according to which the calluses gave suspensions consisting of both the embryogenic and non-embryogenic cluster of the cells in 2, 4- D + Kin (1+1 mg/l) and gave best response (70%). The present study also describes the comparison of *in vivo* and *in vitro* antimicrobial activity of *Vinca rosea*. Methanolic extracts of *in vivo* leaf, *in vitro* leaf callus, *in vitro* leaf, *in vitro* nodal callus and *in vitro* fruit callus were used against three different pathogenic bacterial strains (*Bacillus subtilis, Bacillus licheniformis* & *Azotobacter sp.*) and three different fungal strains (*Aspergillus niger, Alternaria solani* & *Rhizopus oryzae*). Almost all the extract concentrations showed antimicrobial activity indicated by zone of inhibitions (mm). Among all the extracts, maximum zone of inhibition (30.3 mm) was formed by *in vitro* leaf callus extract concentration of 20 mg/ml against *Bacillus subtilis* and *in vitro* leaf extract concentration 2 mg/ml showed 30.0 mm maximum zone of inhibition against *Bacillus licheniformis*. Similarly, maximum zone of inhibition (40.0 mm) was formed by *in vitro* leaf callus extract concentration of 16 mg/ml against *Aspergillus niger*. Present study also reveals that *in vitro* extracts showed better results as compare to the *in vivo* extract for both the antibacterial as well as the antifungal activity.

1

INTRODUCTION

Plants have been an important source of medicine for thousands of years. The WHO estimates that up to 80% of people still believe on traditional remedies such as herbs for their medicines. Plants are the sole source of many modern medicines. The most popular and valuable anti-cancer agents such as paclitaxel and vinblastine are derived solely from plant sources. (Katzung, 1995 & Pezzuto, 1996)

Medicinal plants are the most important source of life saving drugs for the majority of the world's population. Today so many medicinal plants of commercial importance face extinction due to increase in demand and destruction of their habitats due to urbanization and industrialization. (Shrivastava & Singh, 2011)

Among the valuable medicinal plants *Vinca rosea* L. possess an important and significant place in the list of medicinal plants. *Vinca rosea* L. belongs to family *Apocynaceae* is an herbaceous shrub also known as Madagascar periwinkle or *Catharanthus roseus* worldwide. It is cultivated mainly for its alkaloids, which possess anticancerous activities. (Jaleel *et al.*, 2006)

Catharanthus roseus L. or Madagascar periwinkle is one of the most extensively investigated medicinal plants. Since 1950, 1200 scientific publications, including about 85 patents dealing with this plant appeared. (Anonymous Chemical abstract, 1988-1993). Use of plant in medicines has a long history for the treatment of various diseases. The earliest known records for the use of plants of Vinca as drugs are from Mesopotamia in 2600 B.C., and these still are a significant part of traditional medicine and herbal remedies. (Koehn *et al.*, 2005)

There are numerous traditional medicinal plants reported to have hypoglycaemic properties such as *Allium sativum* (Garlic), *Azadirachta indica* (Neem), *Vinca rosea* (Nayantara), *Trigonella foenum* (Fenugreek), *Momordica charantia* (Bitter ground), *Ocimum santum* (Tulsi). (Ahmed *et al.*, 2010)

Catharanthus roseus have been used in folk medicine as an antidiabetic, diuretic, and antidysenteric, an antihemorrhagic and for wound healing. In Europe, it was mainly used as an antidiabetic, for easing lung congestion and throat inflammation, in south and central Africa; as a poultice to stop bleeding in the US; in India in case of insect stings; against eye inflammation in the Caribbean and as an astringent, diuretic and cough remedy in China. (Pahwa, 2009)

Various *in vitro* techniques as micropropagation from existing and adventitious meristems or organ, tissue and cell cultures provide a large amount of *Catharanthus roseus* plant material for the isolation of alkaloids having medicinal properties. *C. roseus* can be successfully regenerated from *in vitro* cultures. The system for high-frequency *C. roseus* plant regeneration by the somatic embryogenesis is also possible. (Pietrosiuk *et al.*, 2007).

A simple, efficient and reproducible regeneration system for *in vitro* propagation of *Catharanthus roseus* via nodal explants cultured on MS medium supplemented with different concentrations of BAP, NAA and IBA was also reported. (Faheem *et al.*, 2011)

Catharanthus roseus is an important antimicrobial plant for novel pharmaceuticals since most of the bacterial pathogens are developing resistance against many of the currently available antimicrobial drugs. The anticancer properties of *Catharanthus roseus* has been the major interest in all investigations. The antimicrobial activity has been checked against different microorganisms. The findings showed that the extracts from the leaves of this plant can be used as prophylactic agent in many of the diseases, which sometime are of the magnitude of an epidemic. (Prajakta *et al.*, 2010)

Catharanthus roseus possesses antibacterial, antifungal, antidiabetic, anticancer and antiviral activities. The extracts have demonstrated significant anticancer activity against numerous cell types. (El-Sayed & Cordell, 1981) It has also been used to treat a wide assortment of diseases including diabetes. The extract of *C. roseus* significantly increased the wound breaking strength in the incision wound model (Nayak & Pereira, 2006). The antibacterial potential in crude extracts of different parts (i.e., leaves, stem, root and flower) of *C. roseus* against clinically significant bacterial strains was reported. (Muhammad *et al.*, 2009)

The inhibition of conidial germination of four fungi such as *Bipolaris sorokiniana, Fusarium oxysporum* f. sp. *vasinfectum, Rhizopus artocarpi* and *Botryodiplodia theobromae* was tested against the extracts of different parts of *Vinca rosea* and *Azadirachta indica* and showed good results in their inhibition. (Alam *et al.*, 2002)

The biotechnological tools are important to select, multiply and conserve the critical genotypes of medicinal plants. Plant tissue culture techniques offer an integrated approach for the production of standardized quality phytopharmaceutical through mass-production of consistent plant material for physiological characterization and analysis of active ingredients. (Debanth 2006)

In conventional cultivation many plants do not germinate, flower and produce seed under certain climatic conditions or have long periods of growth and multiplication. Micropropagation ensures a good regular supply of medicinal plants, using minimum space and time (Prakash & Staden, 2007)

Tissue culturing of medicinal plants is widely used to produce active compounds for herbal and pharmaceutical industries (Sidhu, 2010). *In vitro* propagation of plants holds a tremendous potential for the production of high quality plant based medicines (Murch *et al.*, 2000)

Plant regeneration from shoot and stem meristems has yielded encouraging results in medicinal plants like *Catharanthus roseus, Cinchona ledgeriana* and *Digitalis* spp, *Rehmannia glutinosa, Rauvolfia serpentina, Isoplexis canariensis*10-12. (Paek *et al.*, 1995; Roy *et al.*, 1994 & Perez-Bermudez *et al.*, 2002)

The description for the procedure of development of micropropagation into three different developmental stages: stage I, establishment of explant aseptically; stage II, multiplication of propagules by repeated subcultures on a specific nutrient medium; and stage III, rooting and hardening of plantlets and planting into soil. (Murashige, 1974)

Plant cell culture is a significant tool for plant biochemistry and molecular biology, and available methods include regeneration of differentiated cultures such as the whole plant and organ cultures; shoots, roots and adventitious roots or dedifferentiated cultures such as calluses, cell suspensions and protoplasts. (Endress, 1994)

In past few years, the study of secondary metabolite production in plant cell cultures has received widespread attention. Cell suspension culture is the application of elicitor treatment on plants which can be used to promote culture cells synthesized for the rapid and useful production of secondary metabolites rapidly. (Wu *et al.*, 2002)

Cell suspension cultures are rapidly dividing homogenous suspensions of cells grown in liquid nutrient media from which samples can be taken (King, 1984). In a cell suspension, a mass of cells, called callus, must first be collected. The callus can then be suspended in a liquid callus induction media containing all the required nutrients and elements to allow for optimal growth which acts to turn all cells into undifferentiated cells. The cell suspension is then placed on a shaker to allow the cell aggregates to disperse to form smaller clumps and single cells that are equally distributed throughout the liquid media. The cells will continuously grow until one of the factors becomes limiting causing cell growth to slow. (Pighin, 2003)

Callus, which may be a source of alkaloids, is mostly used for the establishment of cell suspension culture or for induction of shoots or roots. Cell suspension cultures of *Vinca rosea* were studied extensively to obtain highly productive alkaloids, especially vinblastine and vincristine. An artificial method of chemical or enzymatic coupling of the parent monomeric indole alkaloids vindoline and catharanthine to form vinblastine has been successfully used. (Pietrosiuk *et al.*, 2007)

Keeping in view the medicinal value of this plant, plant tissue culture technique was used and evaluated the antimicrobial (antibacterial & antifungal) potential of the plant.

In this study, improvement of *Vinca rosea* was undertaken by applying plant tissue culture techniques for rapid propagation of *V. rosea*. Comparison of antibacterial and antifungal activity of *in vivo* and *in vitro Vinca rosea* plants was also the part of this study. The present research was focused to:

- Micropropagation and callus formation of *Vinca rosea* by using different parameters to get disease free plants.
- Establishment of the cell suspension culture.
- Extraction of alkaloids with the help of methanol was carried out and their comparison of antimicrobial activity both *in vivo* and *in vitro*.

LITERATURE REVIEW

Vinca rosea belongs to family *Apocynaceae*. The common names of the plant are Periwinkle, Sadabahar, Madagascar Periwinkle and Rosy Periwinkle. The plant is of great medicinal significance. Roots and leaves are used in medicines. The plant contains various alkaloids which are anticancerous, antimicrobial, antidiabetic, sedative, hypertensive and tranquilizer in nature.

Ganapathi & Kargi, (2011) reported the significant development in cell culture techniques for production of indole alkaloids from *Catharanthus roseus*. Key points of their work included the effects of nutrients, environmental effects, stress-inducing compounds and strain selection techniques on the production of alkaloids. Cultivation methods such as suspension cultures, immobilization, and a novel biofilm configuration are compared.

Mustafa *et al.,* (2011) described the methods for the initiation, growth and cryopreservation of plant cell suspension cultures. They concluded that *in vitro* dedifferentiated plant cell suspension cultures are better for the large-scale production of fine chemicals in bioreactors and for the study of cellular and molecular processes for the study of plants. Cell suspension cultures contain a relatively homogeneous cell population, allowing rapid and uniform access to nutrition, precursors, growth hormones and signal compounds for the cells. The protocol covers all steps from plant to cell suspension culture, and includes callus initiation from which cell suspension cultures can be obtained.

Aslam *et al.*, (2010) reported *Catharanthus roseus* as an important medicinal plant. Traditionally, different parts of plant were used in the treatments of various diseases such as diabetes, menstrual regulators, hypertension, cancer and antigalactagogue. Moreover, more than 130 alkaloids have been isolated from different parts; by all two important alkaloids (Vinblastine and Vincristine used in cancer treatment) present in very low concentrations. Thus, various *in vitro* biotechnological and biochemical approaches have been used worldwide; which are directly concerned with the *in-vitro* micropropagation and the enhancement of important secondary metabolites present in different parts of *Catharanthus* and being used in the treatment of various diseases.

Aslam *et al.*, (2010) explained the pharmaceutical importance and the low content in the plant of vinblastine and vincristine *Catharanthus roseus* became an important model system for biotechnological studies on plant secondary metabolism. Researchers are focusing their attention to enhance the alkaloids yield by various ways (chemically, enzymatically, synthetically or by cell culture method). The plant cell can be cultured at large scale, but the yield of alkaloids production is too low and limits commercial applications. In recent times, however, two strategies have been commonly used for the enhancement of alkaloids:

a) In vitro cultivation of shoot via organogenesis and somatic embryogenesis, callus or suspension by the optimization of media, phytohormone, temperature, pH, light, aeration etc. In addition, high cell density culture, elicitor's treatment, mutagenesis, bioreactors and immobilization are also practiced to improve alkaloids yield.

b) Genetic engineering and over expression of biosynthetic rate limiting enzymes in alkaloid biosynthesis pathways.

Aslam *et al.*, (2010) explained the antibacterial activity*:* Benzene extract of dried flowers at a concentration of 50% on agar plate was active on *Proteus*, *Pseudomonas*, *Shigella* and *Staphylococcus* species; however, benzene extract of leaves at a concentration of 50% on agar plate was active on *Proteus*, *Pseudomonas*, *Salmonella*, *Shigella* and *Staphylococcus* species. Ethanol (70%) extract of dried leaves on agar plate was active on *Bacillus megaterium* and *Staphylococcus albus* and inactive on *Bacillus cereus* and *Staphylococcus aureus*. Total alkaloids of root at a concentration of 500.0 mg/ml in broth culture were inactive on *E. coli*, *Salmonella lyphosa* and *Shigella dysenteries*. Water extract of entire plant on agar plate at a concentration of 1:4 was inactive on *Salmonella paratyphi*.

Aslam *et al.*, (2010) mentioned the procedure of antifungal activity from leaves and roots on agar plate were active on *Pythium aphanidermatum*. Hot water extract of dried leaves in broth culture was active on *Trichophyton mentagrophytes*. Hot water extract of dried stem in broth culture was active on *T. mentagrophytes* and weakly active on *T. rubrum*. Acetone and water extracts of dried aerial parts at a concentration (50%) on agar plate was inactive on *Neurospora crossa*.

Sidhu, (2010) reported that tissue culturing of medicinal plants is widely used to produce active compounds for herbal and pharmaceutical industries. Conservation of genetic material of many threatened medicinal plants also involves culturing techniques. This work reviews *in vitro* micropropagation techniques and gives examples of various commercially important medicinal plants. The effect of media formulations and culturing techniques on the growth and multiplication of medicinal plants; and on the production of secondary metabolites is also reviewed. Another method of obtaining secondary metabolites is biotransformation.

Karuppusamy, (2009) reported that plant cell and tissue cultures can be established routinely under sterile conditions from different explants such as plant leaves, stems, roots, meristems etc for multiplication and extraction of secondary metabolites. The major advantage of the cell cultures include synthesis of bioactive secondary metabolites, running in controlled environment, independently from climate and soil conditions. The use of *in vitro* plant cell culture for the production of chemicals and pharmaceuticals has made great strides building on advances in plant science.

Taha *et al.,* (2009) reported the production of some pharmaceutical indole alkaloids in *Catharanthus roseus*. Suspension cultures have been induced from leaf explants of *C. roseus* on MS-medium containing Kinetin under light condition. The influence of L-tryptophan; L-glutamine; L-asparginine; L-cystine and L-arginine at the concentrations of 0; 100; 300 or 500 mg/l on enhanced either cell growth characteristics and indole alkaloids production was investigated. The highest value of mass cell cultures and indole alkaloids production were achieved with modified MS medium containing 300 mg/l of either L-glutamine for mass cell induction or L-tryptophan for enhancement and enrichment of total indole alkaloids; vinblastine and vincristine as compared with other used amino acids at different concentrations.

Azimi *et al.,* (2008) explained that *Catharanthus roseus* is one of the most important medicinal and ornamental plants in the world. In this investigation, periwinkle seeds, after sterilization were cultured on MS medium. Petiole segments of seedlings were sub-cultured to medium containing various concentrations of NAA accompanied with Kin and sub-cultured to regenerate the callus and root. Callus and roots were obtained from petioles in some of treatments. The extracts of callus and roots from different treatments were analyzed by spectrophotometer, TLC and HPLC with respect to the indole alkaloids producing capacity. Alkaloids were produced callus and roots from petiole of *C. roseus* in the presence of Kin and NAA. MS medium with 0.1 mg/l NAA + 0.1 mg/l Kin had the highest vindoline, catharanthine, and vincristine and root organogenesis capacity. But the level of these alkaloids and ajmalicine were very low compared to that in petiole of intact plant, and the level of serpentine was similar. New roots, callus roots, and callus from MS medium containing 0.1 mg/l NAA + 0.1 mg/l Kin were sub cultured in hormone-free and 0.1 mg/l NAA + 0.1 mg/l Kin media and for organogenesis and growth. The most alkaloids amount was produced in new roots and callus roots.

Goyal *et al.,* (2008) reported *Catharanthus roseus* (periwinkle) as an important medicinal plant. The plant was selected to evaluate the possibility for novel pharmaceuticals since most of the bacterial pathogens are developing resistance against currently available antibiotics. Extraction of each plant part in appropriate solvent followed by evaluation of antibacterial activity by agar well diffusion assay against a total of six bacterial strains. Further, minimum inhibitory concentration(s) was evaluated for active crude extracts. Data indicated that the pattern of inhibition depends largely upon the extraction procedure, the plant part used for extraction, state of plant part (fresh or dry), solvent used for extraction and the microorganism tested. Dry powder extracts of all plant parts demonstrated more antibacterial activity than extracts prepared from fresh parts. Furthermore, extracts prepared from leaves were shown to have better efficacy than stem, root, and flower extracts. Organic extracts provided more potent antibacterial activity as compared to aqueous extracts. Among all the extracts, the ethanol extract was found to be most active against almost all the bacterial species tested. Hot water and cold water extracts were completely inactive. Gram-positive bacteria were found more sensitive than Gram-negative bacteria. The study promises an interesting future for designing potentially active antibacterial agents from *Catharanthus roseus.*

Ramani & Jayabaskaran, (2008) established a cell suspension culture protocol from the leaf explants of *V. rosea* and gave the characteristics of *V. rosea* suspension cultured cells.

Liang, (2007) compared the use of plant *in vitro* culture techniques with the classical methods of *in vivo* vegetative propagation. *In vitro* cloning has been proven to be an important tool in speeding up propagation. *In vitro* propagated plants are often healthier than those clones *in vivo*. This is mainly due to rejuvenation and they are often disease-free plants. Cell suspension cultures have also been proven to be suitable for continuous production of bio chemicals. The cultured cells are also the material choice for biochemical and molecular investigation of plant secondary metabolites. And scaling up from flaks to bioreactor for the production of phytochemicals is always performed using suspension culture.

Pietrosiuk *et al.*, (2007) reported the various methods of *in vitro* culture of *Catharanthus roseus* provide new sources of plant material for the production of secondary metabolites such as indole alkaloids. Callus, cell suspension, plantlets, and transgenic roots cultured in the bioreactor are used in those experiments. The most promising outcomes include the production of the following indole alkaloids: ajmalicine in unorganised tissue, catharanthine in the leaf and cell culture in the shake flask and airlift bioreactor, and vinblastine in shoots and transformed roots. The method of catharanthine and ajmalicine production in the suspension culture in bioreactors has been successful. The transformed root culture seems to be the most promising for alkaloid production. Transformed hairy roots have been also used for encapsulation in calcium alginate to form artificial seeds.

Batra *et al.*, (2006) reported the rapid expansion of phytopharmaceutical industries and the ever increasing demand for natural resources of important life-saving plant-based drugs have placed great pressure on the natural existence of a number of economically important plants, otherwise known as medicinal plants.

Shariff *et al.*, (2006) reported the *in vitro* antimicrobial activity of *Rauvolfia tetraphylla* and *Physalis minima* leaf and callus extracts which were studied against selected pathogenic fungi and bacteria, following broth dilution assay. Leaves and calli were extracted using absolute alcohol, benzene, chloroform, methanol and petroleum ether. Among the five solvents used, leaf and callus extracted in chloroform of both the plants were found to be more effective against pathogenic bacteria and fungi, where the minimum inhibitory concentration (MIC) ranged between 0.25 to 6 mg/ml. Absolute alcohol extracts showed MIC of 0.25 to 4 mg/ml for bacteria, whereas for fungi it ranged from 0.25 to 100 mg/ml. Extracts of benzene and petroleum ether were ineffective in inhibiting the bacterial and fungal growth or showed poor inhibition. Methanol extract showed MIC of 0.25 to 100 mg/ml against bacterial pathogens and 0.5 to 100 mg/ml against fungal pathogens.

Gaines, (2004) revealed that a large population relies on pharmaceuticals derived from plants. It is estimated that 75% of the world's population is dependant upon plant derived pharmaceuticals. The need for methods of increasing the production of plant-derived pharmaceuticals cost-effectively and with environmental consideration is becoming more important. Of particular interest are the pharmaceutically valuable alkaloids from the *Catharanthus roseus.*

Yarnell, (2004) gave a quick botanical review about *Catharanthus roseus* i.e., *C .roseus* originates from Madagascar, but has now been spread throughout the tropics and subtropics by human activity. It has readily naturalized to almost every hot climate in which it has been planted. When it was discovered by Europeans, it was wrongly classed as a *Vinca* or true periwinkle. This error was ultimately realized and corrected, and the plant was put in its own genus. Though its primary traditional use was for people with diabetes, *Catharanthus* also has anticancer effects.

Ten-Hoopen *et al.*, (2002) showed that temperature has an important influence on growth and ajmalicine production by *C. roseus* suspension cultures. The optimal temperature for both processes i.e. biomass growth and secondary metabolite production was 27.5° C.

Andrews, (2001) explained the biological screening of the plant which was carried out for the determination of antimicrobial activity both *in vivo* and *in vitro*. The well plate method was used to determine the antimicrobial activity of *Catharanthus roseus,* for this reason wells were made on petri plates filled with nutrient agar containing a culture of a particular bacterial species. The plant extract was added and after the incubation period, the zones of inhibitions were measured indicating the antibacterial activity of the plant extracts against different gram positive and gram negative microorganisms.

Zheng & Wang, (2001) showed an *in vitro* analysis in which *Catharanthus* was the most potent antioxidant herb analyzed among many others, including *Thymus, Salvia* and *Rosmarinus.* This was moderately correlated to the high phenolic content in the plant.

Datta & Srivastava, (1997) reported the production of vinblastine, an anticancer agent, by *Catharanthus roseus,* increased as the seedlings matured, attaining a steady concentration after the plants becomes more than three month old. Vinblastine could be detected in the callus lines established from different explants. As the callus differentiated multiple shoots, the vinblastine production increased rapidly, equalizing to that of *in vivo* seedlings of similar age. The high degree of differentiation and maturity in the tissues of *Catharanthus* was correlative to the increased vinblastine production, both *in vivo* and *in vitro*. Increased production of vinblastine, an anticancer agent, was correlative to the degree of differentiation and maturity in *Catharanthus roseus* during *in vivo* and *in vitro* regeneration from callus.

Nisbet & Moore, (1997) reported that the global attention has been shifted towards finding new chemicals, specifically herbals, for the development of new drugs. These natural products can provide unique elements of molecular diversity and biological functionality, which is indispensable for novel drug discovery.

Moreno *et al.,* (1996) studied the effect of elicitation on different metabolic pathways involved in the secondary metabolism of *V. rosea* cell suspension cultures.

Shah & Chauhan, (1996) showed that tissue culture technique is developed for the development of dimeric alkaloids. They reported a detailed application of *C. roseus* including traditional uses in various developed and developing countries, pharmacological activities and the application of various biotechnological tools.

Gamborg & Phillips, (1995) explained about cell suspension culture. According to them a plant cell suspension culture is a sterile (closed) system normally initiated by aseptically placing friable callus fragments into a suitable sterile liquid medium.

Fulzele & Heble (1994) and Ten-Hoopen *et al.*, (1994) performed a larger-scale production of catharanthine and ajmalicine in suspension cultures of *V. rosea*, in bioreactors of different volumes and types.

Kim *et al.*, (1994) described the system for high frequency *C. roseus* plant regeneration by the somatic embryogenesis. The authors obtained the callus from the anthers of *C. roseus* plants using the solid MS medium supplemented with NAA 1.0 mg/l and kinetin 0.1 mg/l. The somatic embryos formed following transferring of the callus into the liquid MS medium supplemented with NAA and kinetin. The plants thus grown had the same number of chromosomes as the plants grown from the seeds (2n = 16).

Yuan & Hu, (1994) investigated the influence of different combinations of auxins, cytokinins and light intensity on the formation of multiple shoots of *C. roseus* in *in vitro* cultures. The authors demonstrated that the growth regulators BAP and NAA added to the MS medium, 7.0 mg/l and 1.0 mg/l, respectively, potently stimulated the formation of shoots, whereas 2, 4-D suppressed their differentiation and formation. The light intensity of 550–700 lux was found to be beneficial with the simultaneous use of MS medium supplemented with BAP 2 mg/l and NAA 0–1.0 mg/l.

Marfori & Alejar, (1993) investigated the alkaloid productivity of the callus tissue derived from roots, stems, leaves and flowers of two *C. roseus* varieties, white and pinkie purple. The authors obtained 8 callus lines using the Linsmaier-Skoog (LS) medium supplemented with BAP 3.0 mg/l, 2, 4-D 0.5 mg/l, and coconut milk. Then the tissue formed was transferred into the LS medium without growth regulators and cultivated for 1 year. The alkaloid content was higher in the callus derived from the pinkie purple periwinkle variety and in the callus derived from roots.

Sadowska, (1991) proposed the climatic conditions and soil properties of some European countries which were unfavourable for the cultivation of *C. roseus*. It may be cultivated only as an annual plant in greenhouses and plastic tunnels but then the content of dimeric indole alkaloids is very low.

Perez *et al.*, (1990) used various parts of *C. roseus* (leaf, stem, flower and root) and the extracts of these different parts were subjected to antibacterial assay. The extracts of *C. roseus* did not exhibit antibacterial activity against *Staphylococcus aureus*. Moreover, leaf, stem and flower extracts were also ineffective against *Pseudomonas aeruginosa*. The leaf extract did not exhibit activity against *Corynebacterium diphtheriae*; similarly, the crude extract of stem did not shown activity against *Shigella boydii*. The most effective was the root extract, which exhibited broad-spectrum antibacterial activity against *Salmonella typhimurium* and *S. boydii*. The flower extract showed activity against *C. diphtheria*.

Heijden *et al.*, (1989) regenerated *Catharanthus roseus* successfully from *in vitro* cultures using various techniques, e.g., micropropagation using existing and adventitious meristems. Generally, plant regeneration was accomplished using somatic organogenesis. Shoots formed and grew out of the callus tissue. Studies conducted in different highly specialized laboratories have demonstrated that in many cases the ability to produce secondary metabolites is associated with the process of organogenesis. This finding has contributed to considerable progress in the field of micropropagation.

Hirata *et al.*, (1987) showed the multiple shoot cultures of *C. roseus* were directly induced, with high frequencies, from seedlings on the MS medium supplemented with BAP 1.0 mg/l. The resulting cultures also consisted of unorganized tissue directly on the solid medium and multiple shoots having several small leaves.

Morris, (1986) established the callus culture from leaves of *C. roseus* using three media: Gamborg (B5), MS and Zenk's medium in different modifications (B5 with 2,4-D 1.0 mg/l and kinetin 0.1 mg/l; B5 with IAA 1.0 mg/l and kinetin 0.1 mg/l; MS with NAA 1.0 mg/l and kinetin 0.1 mg/l, Zenk's medium supplemented with IAA 0.175 mg/l and BAP 1, 125 mg/l, and also with NAA, 2,4-D and zeatin at 1.0 mg/l). The culture conditions of *C. roseus* callus tissue are very important for alkaloid production.

Takeuchi & Komamine, (1982) prepared the protoplast from suspension-cultured *Vinca rosea* cells were grown in a liquid medium, and the effects of osmolarity and growth regulators on cell division and on the composition of regenerated cell walls were investigated. The concentration of mannitol optimal for cell division was 0.3-0.4 M. The presence of 2, 4-D was essential for cell division, and BAP enhanced cell division at concentrations of 0.03- 0.1 ppm. However, the composition of regenerated cell walls was abnormal under suspension culture; the predominant sugar was glucose, indicating that the regenerated cell walls consisted mostly of glucans, and that the other cell-wall components were released into the medium. Mannitol, 2, 4-D, and BAP at various concentrations did not significantly affect the sugar composition of the regenerated cell walls. Compared with liquid culture, cell division was stimulated when protoplasts were cultured on agar, and their regenerated cell walls had a composition similar to that of the original culture.

MATERIAL AND METHODS

Present investigation deals with *in vitro* micropropagation, callogenesis, cell suspension culture and antimicrobial activity of *Vinca rosea.*

A. Source of Explants:

The plants used as explants were taken from Botanical Garden of Lahore College for Women University, Lahore and from Green House of Lawrence Garden, Lahore.

B. Methodology:

The following methodology was adopted for this work.

3.1 Sterilization:

In tissue culture technique, one of the vital steps is the satisfactory and reliable sterilization of glassware, growth media and explants.

3.1.1 Sterilization of Glassware:
The glassware was cleaned and sterilized in the following way:-

- All the glassware (test tubes, pipettes, beakers, jars, cylinders, funnels, stirrers etc.) were washed thoroughly with detergent and were rinsed under tap water several times.
- Glassware was then soaked in chromic acid ($K_2Cr_2O_7+H_2SO_4+H_2O$) for 10 hrs. After soaking in acid, this glassware was washed thoroughly with running tap water and with double distilled water.
- Finally glassware was sterilized in an autoclave at 121^0C for 20 min at 15 lb/inches^2.

3.1.2 Sterilization of Instruments:
Sterilization of forceps and scalpels was done by soaking the instruments in 95% ethanol and flaming off the instruments before use.

3.1.3 Media Sterilization:

The culture vessels containing media must be sterilized to avoid contamination which is a very serious problem in plant tissue culture technique. The test tubes were wrapped with plastic sheets which we called polyethylene bags and tied with rubber bands. Then bunches of test tubes were wrapped with brown paper and autoclaved at 121^0C for 20 min at 15 lb/inches^2. Each vessel was marked to show the precise composition of medium and also the date of preparation.

3.1.4 Storage of Media:

The medium was stored at temperature of 22 ± 2^0C after autoclaving.

3.1.5 Surface Sterilization of Explants:

The explants were surface sterilized for successful growth of explants and successful callus formation. The field collected material was washed several times with tap water and detergent first.

The *Vinca rosea* explants were then surface sterilized with the bleach (40%) for 20 min. Then several rinses were given with autoclaved water till the removal of smell of sodium hypochlorite.

In dealing with *Vinca rosea*, following explants were used for micropropagation and callogenesis:

1. Shoot tips
2. Nodal portion
3. Leaf section
4. Fruit

3.1.6 Sterilization of Laminar Air Flow Cabinet:

Following essential steps were taken for sterilization of laminar air flow cabinet before starting the work:

- Cleaned the surface of laminar flow cabinet with ethanol.
- Irradiated with UV light for about half an hour before work.
- Turned on the fluorescent light and blower during work.

3.1.7 Culture room Sterilization:

Sterilization of culture room was done on the daily basis. The culture room was sprayed thoroughly with 90% v/v ethanol by using spray gun. Formaline was also kept in the culture room for at least 1-2 days in a month to minimize the rate of contamination.

3.2 Concentrations used for preparing the Stock Solution:

Stock solutions consisted of the following group of chemicals with their strength:

1. Macronutrients   (20X)
2. Micronutrients   (100X)
3. Vitamins   (100X)
4. Iron-EDTA   (200X)
5. Growth hormones (Auxins, Cytokinins, Gibberellins)

Formulation of these nutrients was done according to MS medium and finalized by adding gelling agent (phytagel) in it. (Murashige and Skoog, 1962)

3.3 Preparation of Media:

- To prepare 1 litre of MS medium, following volumes of stock solutions were used which were measured carefully by using the correctly sized pipette or graduated cylinder and the proper range balance.
- 50 ml of macronutrient stock solution having 20X was added.
- 10 ml of micronutrient stock solution having 100X was used.
- 10 ml of vitamin stock solution having 100X was added.
- 5 ml of Fe-EDTA stock solution having 200X was added.
- Growth regulators were added according to the requirement.
- Sucrose was added to the medium at 3% concentration (30g/l).
- Final volume was made up to 1000 ml by adding double distilled water.
- After mixing well, pH of the medium was adjusted to 5.5-5.7 using 0.1N KOH or 0.1N HCl.
- For solid medium 1.5 g/l phytagel was added and was mixed thoroughly.
- For liquid medium, cotton was poured in jars without addition of phytagel.

- Volume of this medium was dispensed into pre-sterilized culture tubes and tubes were covered with polypropylene sheets and then wrapped with the help of rubber bands.
- The test tubes were labelled with the name of the medium and date of preparation.
- Medium was sterilized by autoclaving at $121^{\circ}C$ for 20 min. at 15 lb/inches^2.
- After autoclaving culture tubes were cooled down to room temperature and were stored at $22 \pm 2^{\circ}C$ till use.

3.4 Conditions for Inoculation of Explant:

For aseptic transfer of explants following steps were followed:
- Hands were washed with soap and water.
- Laminar air flow cabinet was cleaned with ethanol.
- Turned on the UV light 15-30 min before starting.
- Quickly wrapped the containers with covers after the completion of operational step.
- Talking in the cabinet was avoided, cover mouth with surgical mask.
- Wear hand gloves and cap during the work.
- When finished, turned off the burner and all materials were removed.
- Surface was cleaned with ethanol.
- Turned off the laminar air flow cabinet after working.

3.5 Culture room environment:

- **Temperature:**

Optimum temperature required for culture environment was maintained at $22 \pm 2^{\circ}C$.

- **Photoperiod:**

The cultures were provided with 16 hrs. light period (from cool white florescent tubes) and 8 hrs. dark with light intensity of 2000-3000 lux.

3.6 Plan of experiment and data recording:

The cultured explants were observed after inoculation and the data was recorded about the contamination, percentage of shoot formation, callus formation and number or frequency of regenerated plants per explants.

3.7 Cell suspension culture of Vinca rosea:

Cell suspension was initiated by taking a friable callus formed in the test tube and then dividing it into fragments in sterile environment. Then the small pieces of the callus were placed in the 250 ml shake flasks and then the flasks were placed on the orbital shaker at 100 rpm.

3.8 Statistical analysis:

All experiments were arranged in a Completely Randomized Design and data were analyzed using the Costat software. Data (mean ± SD) were collected from five experiments each with (3-5) replicates based on Duncan's new multiple Range Test (DMRT).

3.9 Equipments and Facilities:

All the work was carried out under aseptic conditions. To achieve this purpose special equipments and facilities are required which are following

1. Glassware (flasks, beakers, cylinders, pipettes, stirrer, test tubes and Petri plates etc.)
2. Oven for dry sterilization
3. Autoclave
4. Laminar air flow cabinet for sterile transfer
5. Water distillation unit
6. Refrigerator
7. Microwave oven
8. Electric balance
9. pH meter
10. Incubator
11. Orbital shaker
12. Light and temperature controlled growth incubator
13. Cellophane paper, rubber bands and newspapers
14. Gas burner
15. Stainless steel scissor, fine forceps, scalpel, spatula etc.

C. Antimicrobial Activity:

3.1 Plant Collection:

The *Vinca rosea* plants used as explants were taken from Botanical Garden of L.C.W.U., Lahore and from Green House of Lawrence Garden, Lahore.

The following explants were used for antimicrobial activity:

1. *In vivo V. rosea* leaves
2. *In vitro V. rosea* leaves
3. Leaf callus
4. Nodal callus
5. Fruit callus

3.2 Extraction Procedure:

3.2.1 Soxhlet extraction:

Soxhlet apparatus was used for the extraction of alkaloids from *in vivo V. rosea* leaves. The weighed fresh *in vivo* plant material was crushed in pestle and mortar first and then placed in the extraction thimble. The weighed amount was placed in an extraction chamber which was suspended above the flask containing the solvent methanol and below a condenser. The flask was heated and the methanol evaporated and moved into the condenser where it was converted into a liquid that trickled into the extraction chamber containing the plant material. The extraction chamber was designed so that when the solvent surrounding the sample exceeded a certain level it flowed and trickled back down into the boiling flask. At the end of the extraction process, the flask containing the methanol extract was removed and methanol was evaporated by using rotary evaporator. The weight of the extract was measured and percentage yield of the plant material was calculated.

3.2.2 Extraction by Filtration Method:

Alkaloids from *in vitro V. rosea* leaves, leaf callus, nodal callus and fruit callus were extracted by simple filtration method. The weighed amount of each sample was first crushed separately in pestle and mortar and then they were filtered in respective flasks.

3.3 Microorganisms:

Three different strains i.e. *Bacillus subtilis, Bacillus licheniformes* (gram positive bacteria) and *Azotobacter sp.* (gram negative bacteria) were used for testing antibacterial activity. Similarly three different strains i.e. *Aspergillus niger, Alternaria solani* and *Rhizopus oryzae* were used for testing antifungal activity. The test organisms used in this study were obtained from Government College University (GCU), Lahore, Pakistan. The microorganisms were cultured on nutrient agar slants. The cultures were maintained by sub culturing periodically and preserved at 4°C prior to use.

3.4 Assay for Antimicrobial Activity:

Antimicrobial activity was tested by agar well diffusion method (Mukherjee *et al.*, 1995). Different concentrations of the Vinca alkaloids were prepared in organic solvent i.e. methanol by using serial dilution method. 0.5ml of the 24h fresh cultures were poured into 35ml sterile molten nutrient agar in sterile petri plates and allowed to be solidified. After solidification, 7 mm wells were made using sterile borer. The wells were then filled with 0.1ml of the sample extract. The antibacterial assay plates were incubated at 37°C for 24h and antifungal assay plates were incubated at 25°C for 48h. After incubation, the zones of inhibition were measured. Each experiment was carried out in triplicate and mean diameter of the inhibition zone was recorded.

3.5 Minimum Inhibitory Concentration (MIC):

The extract which showed antimicrobial activity in agar well assay was subjected to MIC assay (Jones *et al.*, 1985). In order to determine MIC, serial dilutions of the extract were prepared with concentration ranged from 2 mg/ml to 20 mg/ml. Minimum inhibitory concentration (MIC), which was determined as the lowest concentration of plant extracts inhibiting the growth of the organism, was determined based on the readings. All tests were performed in triplicate.

RESULTS & DISCUSSION

The present study deals with the micropropagation, callogenesis and establishment of cell suspension cultures. The present study is also concerned with comparison of *in vivo* and *in vitro* antimicrobial activity of Vinca rosea.

Results

Micropropagation:

Different concentrations of cytokinins alone and their combinations with auxins were used for micropropagation of shoot tips and nodal portion of *Vinca rosea*.

4.1: Effect of BAP alone or in combination with the NAA on shoot induction from shoot tip explants of *Vinca rosea:*

Table 4.1 shows the effect of BAP alone or in combination with NAA on shoot induction from shoot tip explants of *Vinca rosea.* MS basal medium without growth regulators served as the control. The response of this medium to shoot induction was only 35%. The medium supplemented with BAP alone gave better response towards shoot induction as compare to their association with NAA. The response of shoot initiation of medium supplemented with BAP alone at various concentrations (1.0 mg/l, 2 mg/l & 3 mg/l) was 90%, 70% & 60% respectively. On the other hand, medium supplemented with different combinations of BAP+NAA (1+1 mg/l, 1+2 mg/l & 2+1 mg/l) showed delayed response with the percentage of 70%, 70% & 50% respectively.

Similarly, the basal medium showed delayed shoot initiation in approximately 21 days where as various concentrations i.e., 1 mg/l, 2 mg/l and 3 mg/l of BAP alone took 12, 19 and 23 days for shoot initiation respectively from the shoot tip explants. The combined effect of BAP+NAA with concentrations of 1+1 mg/l, 1+2 mg/l and 2+1 mg/l started shoot initiation in 17, 21 and approximately 25 days of inoculation.

The shoot tip explants inoculated in BAP (1.0 mg/l) showed maximum shoot cultured (10.6±0.55) with the mean length of 5.14cm. Number of shoot cultured and their length (cm) decreases with the increase in BAP concentration.

In combination with NAA, BAP did not show any promising results and both the number of shoot cultured and shoot length was less as compare to BAP alone. BAP alone with 1mg/l showed better results overall.

Plate 4.1 (a, b, c, d, e, f, g, h & i) showed the effect of BAP alone or in combination with the NAA on shoot induction from shoot tip explants of *Vinca rosea*.

Table 4.1: Effect of BAP alone or in combination with the NAA on shoot induction from shoot tip explants of *Vinca rosea*.

S. No.	Treatment	Media	Growth hormone (mg/l)	Concentration (mg/l)	Response (%)	Days required for shoot initiation	No. of shoots cultured	Average length of shoots (cm)
1	T1	MS	Control (Basal)	-	35	20.4 ± 1.14^a	3.22 ± 0.15^b	2.3 ± 0.19^b
2	T2		BAP	1.0	90	12.0 ± 1.22^c	10.6 ± 0.55^a	5.14 ± 0.18^a
3	T3		BAP	2.0	70	19.6 ± 0.89^b	8.0 ± 1.00^b	4.02 ± 0.13^b
4	T4			3.0	60	23.0 ± 1.00^a	4.4 ± 0.55^c	3.22 ± 0.08^c
LSD	-	-	-	-	-	**1.4453**	**1.0064**	**0.1899**
5	T5	MS	BAP+NAA	1+1	70	17.4 ± 0.55^c	7.8 ± 1.09^a	3.46 ± 0.09^a
6	T6		BAP+NAA	1+2	70	21.0 ± 1.22^b	5.8 ± 1.09^b	2.98 ± 0.08^b
7	T7			2+1	50	24.6 ± 1.14^a	3.8 ± 0.84^c	2.64 ± 0.11^c
LSD	-	-	-	-	-	**1.4007**	**1.4007**	**0.1331**

No. of explant inoculated = 10

Each value is the mean of three replicate with standard deviation (mean±S.D)

Mean followed by different letters in superscript in the same column differ significantly at P=0.05 according to Duncan's new multiple range test.

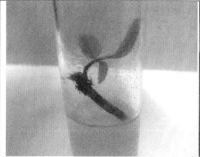

a: Shoot induction in *V. rosea* from shoot tip in BAP 1 mg/l.

b: Multiple shoot formation in *V. rosea* on BAP 1 mg/l from shoot tip explant.

c: Shoot formation in *V. rosea* from shoot tip in the MS basal medium.

d: Multiple shoot formation in BAP 1 mg/l from shoot tip explant of *V. rosea*.

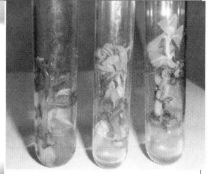

BAP 3 mg/l   BAP 2 mg/l   BAP 1 mg/l

e: Multiple shoot formation in BAP+NAA (1+1 mg/l) from the shoot tip of *V. rosea*.

f: Effect of BAP alone on the shoot proliferation from shoot tip explants of *V. rosea*.

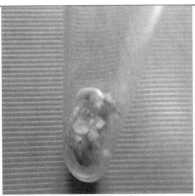

| g: Shoot formation on BAP+NAA (1+1 mg/l) from the shoot tip of *V. rosea*. | h: Shoots cultured on BAP 1 mg/l from the shoot tip of *V. rosea*. |

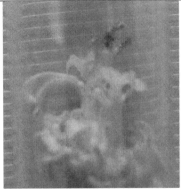

i: Shoots cultured on MS basal medium from the shoot tip of *V. rosea*.

Plate No. 4.1 (a, b, c, d, e, f, g, h & i): **Effect of BAP alone or in combination with the NAA on shoot induction from shoot tip explants of *Vinca rosea*.**

4.2: Effect of BAP alone or in combination with the NAA on shoot induction from nodal explants of *Vinca rosea:*

The micropropagation response of *Vinca rosea* from the nodal explants has been given in the table 4.2. The effect of BAP alone or in combination with the NAA on shoot induction was noticed. It was observed that the basal medium and the medium in combination with BAP+NAA showed delayed response towards shoot initiation as well as the number of shoots cultured and the mean shoot length (cm) was less when compared with the response of the nodal explants inoculated in the medium fortified with BAP alone with different concentrations (1, 2 & 3 mg/l). The comparative analysis of the table 4.2 also revealed that shoot proliferation was most effective in BAP 1.0 mg/l concentration. The response in BAP 1.0 mg/l was maximum i.e. 80%. The shoot initiation in BAP 1.0 mg/l started within 12.8 ± 1.09 days and the number of shoots cultured was maximum (7.8 ± 1.09) with the mean length of 4.3 cm.

The effect of BAP alone or in combination with the NAA on shoot induction from nodal explants of *Vinca rosea* can be clearly seen in the plate number 4.2 (a, b, c, d, e, & f). Plate number 4.3 shows a well acclimatized plant of *V. rosea* developed from nodal portion.

Table 4.2: Effect of BAP alone or in combination with the NAA on shoot induction from nodal explants of *Vinca rosea*.

S. No.	Treatment	Media	Growth hormone (mg/l)	Concentration (mg/l)	Response (%)	Days required for shoot initiation	No. of shoots cultured	Average length of shoots (cm)
1	T8	MS	Control (Basal)	-	30	20.4 ± 0.55^a	3.2 ± 0.45^a	2.4 ± 0.12^a
2	T9	MS	BAP	1.0	80	12.8 ± 1.09^b	7.8 ± 1.09^a	4.3 ± 0.14^a
3	T10		BAP	2.0	60	14.2 ± 1.48^b	6.4 ± 1.34^a	2.4 ± 0.2^b
4	T11			3.0	50	19.2 ± 0.84^a	3.6 ± 0.55^b	2.08 ± 0.08^c
LSD	-	-	-	-	-	**1.6109**	**1.4453**	**0.1779**
5	T12	MS	BAP+NAA	1+1	55	18 ± 0.71^c	5.8 ± 1.09^a	3.84 ± 0.05^a
6	T13	MS		1+2	45	21.2 ± 0.84^b	4.4 ± 0.55^b	3.0 ± 0.07^b
7	T14			2+1	40	25.4 ± 0.89^a	3.2 ± 0.84^c	2.76 ± 0.17^c
LSD	-	-	-	-	-	**1.1251**	**1.1800**	**0.1832**

No. of explant inoculated = 10

Each value is the mean of three replicate with standard deviation (mean±S.D)

Mean followed by different letters in the same column differ significantly at P=0.05 according to Duncan's new multiple range test.

a: Shoot initiation in *V. rosea* from the nodal explant in BAP 1 mg/l.

b: Shoot induction in *V. rosea* from nodal portion in the MS basal medium.

c: Early shoot multiplication in *Vinca* from the nodal explant on BAP 1mg/l.

d: Shoots cultured from the nodal explant in BAP 1 mg/l in *V. rosea*.

e: Multiple shoot initiation in *V. rosea* from the nodal explant in BAP+NAA (1+1 mg/l).

f: Shoots cultured from the nodal explant in MS basal medium in *V. rosea*.

Plate No. 4.2 (a, b, c, d, e & f): **Effect of BAP alone or in combination with the NAA on shoot induction from nodal explants of *Vinca rosea*.**

Plate No. 4.3: **A well acclimatized plant of *Vinca rosea.***

32

Callogenesis:

Different concentrations of auxins alone and in combination with cytokinins were used for the callogenesis of different explants of *Vinca rosea*.

4.3: Effect of auxins singly or in combination with the cytokinins on the growth and nature of leaf callus in *Vinca rosea:*

Leaf tissues were inoculated on MS basal medium or supplemented with cytokinins alone or in combination with auxins to observe response of media towards callus initiation as well as their growth and nature. Depending upon the concentration and combination of hormones used a wide range in frequency of callus formation and nature of callus was observed.

Highest response of callus induction (95%) was observed in the media supplemented with 1+1 mg/l of 2, 4-D+Kin. It is evident from the table 4.3 that the combination of auxins and cytokinins proved very effective in callus induction. The time for callus initiation ranged from 10 to 16 days of inoculation. Delayed callus formation was noticed in the medium supplemented with 2, 4-D+Kin (2+1 mg/l) as well as 2, 4-D (1.5 mg/l).

As shown in the table 4.3, auxins in combination with cytokinins (2, 4-D+Kin) in concentration of 1+1 mg/l showed best results in rapid and early callus formation. The nature of callus formed is light brown and less crystalline.

The effect of auxins singly or in combination with the cytokinins on the growth and nature of leaf callus in *Vinca rosea* can be seen in the plate number 4.4 (a, b, c & d).

33

Table 4.3: Effect of auxins singly or in combination with the cytokinins on the growth and nature of leaf callus in *Vinca rosea*.

S. No.	Treatment	Media	Growth hormone (mg/l)	Concentration (mg/l)	Response (%)	Days required for callus initiation	Callus proliferation	Nature of callus
1	T15	MS	Control (Basal)	-	10	13.2 ± 0.45^a	V. Low	Green and soft
2	T16		2,4-D	1.5	30	15.0 ± 0.70^b	Low	Translucent and hard
3	T17			2.0	40	13.2 ± 1.30^b	Low	Yellowish brown and hard
4	T18			3.0	60	11.6 ± 0.55^c	Moderate	Brown and hard
LSD	-	-	-	-	-	**1.1015**	-	-
5	T19	MS	2,4-D + Kin	1+1	95	10.2 ± 0.45^c	High	Light brown and less crystalline
6	T20			1+2	75	13.4 ± 0.55^b	Moderate	Creamish Brown and fragile
7	T21			2+1	70	16.0 ± 1.41^a	Moderate	Brown and crystalline
LSD	-	-	-	-	-	**1.2579**	-	-

No. of explant inoculated = 10.
three replicate with standard deviation (mean±S.D). Mean followed by different letters in the same column differ significantly at P=0.05 according to Duncan's new multiple range test.

Each value is the mean of

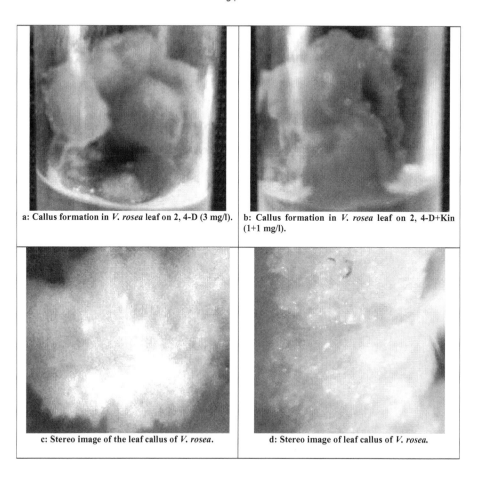

a: Callus formation in *V. rosea* leaf on 2, 4-D (3 mg/l).	b: Callus formation in *V. rosea* leaf on 2, 4-D+Kin (1+1 mg/l).
c: Stereo image of the leaf callus of *V. rosea*.	d: Stereo image of leaf callus of *V. rosea*.

Plate No. 4.4 (a, b, c & d): **Effect of auxins singly or in combination with the cytokinins on the growth and nature of leaf callus in *Vinca rosea.***

4.4: Effect of auxins singly or in combination with the cytokinins on the growth and nature of nodal callus in *Vinca rosea:*

Nodal portion was cultured on MS basal medium or supplemented with cytokinins alone or in combination with auxins to observe response of media towards callus initiation as well as their growth and nature. Table 4.4 show a detailed analysis of the effect of auxins (2, 4-D) alone or in combination with Kinetin (Kin). The table 4.4 revealed that the media supplemented with 2, 4-D+Kin (1+1 mg/l) supported 80% callus induction after 12 days of inoculation which is the best of all the other media mentioned in the table 4.4. The intensity of callus formation was high in the same media and the callus formed was greenish and hard. The growth and nature of the nodal callus formed can be seen in the plate number 4.5 (a, b, c & d).

Table 4.4: Effect of auxins singly or in combination with the cytokinins on the growth and nature of nodal callus in *Vinca rosea*.

S. No.	Treatment	Media	Growth hormone (mg/l)	Concentration (mg/l)	Response (%)	Days required for callus initiation	Callus proliferation	Nature of callus
1	T22		Control (Basal)	-	10	13.8 ± 0.45^b	V. Low	Green and soft
2	T23	MS	2,4-D	1.5	40	18.6 ± 0.89^a	Low	White and Hard
3	T24	MS	2,4-D	2.0	50	17.8 ± 1.30^a	Moderate	White and hard
4	T25			3.0	60	14.0 ± 1.58^b	Moderate	Brown and hard
LSD	-	-	-	-	-	**1.5287**	-	-
5	T26		-	1+1	80	12.0 ± 0.71^b	High	Greenish and hard
6	T27	MS	2,4-D + Kin	1+2	70	12.6 ± 0.55^b	Moderate	Creamish Brown and hard
7	T28			2+1	60	16.0 ± 1.00^a	Moderate	Greenish brown and hard
LSD	-	-	-	-	-	**1.0674**	-	-

No. of explant inoculated = 10

Each value is the mean of three replicate with standard deviation (mean±S.D). Mean followed by different letters in the same column differ significantly at P=0.05 according to Duncan's new multiple range test.

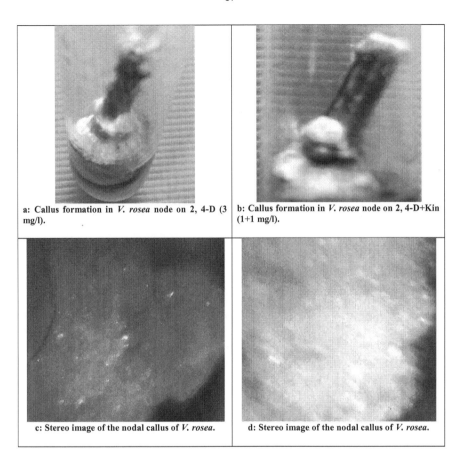

a: Callus formation in *V. rosea* node on 2, 4-D (3 mg/l).

b: Callus formation in *V. rosea* node on 2, 4-D+Kin (1+1 mg/l).

c: Stereo image of the nodal callus of *V. rosea.*

d: Stereo image of the nodal callus of *V. rosea.*

Plate No. 4.5 (a, b, c & d): **Effect of auxins singly or in combination with the cytokinins on the growth and nature of nodal callus in *Vinca rosea.***

4.5: Effect of auxins singly or in combination with the cytokinins on the growth and nature of fruit callus in *Vinca rosea:*

Fruit of Vinca rosea was also cultured on different media for the callus formation. The table 4.5 revealed that the intensity of callus formation was moderate in 2, 4-D+Kin (1+1 mg/l), less in 2, 4-D+Kin (1+2 mg/l) as well as in 2, 4-D (3 mg/l) and very low in the basal medium. The response was maximum (60%) in 2, 4-D+Kin (1+1 mg/l) followed by 40% in 2, 4-D+Kin (1+2 mg/l) and 2, 4-D (3 mg/l) and then 20% in the basal medium. The other concentrations and combinations did not show any response. The callus formed varies from white colour and hard textured to Creamish brown and fragile textured callus.

The plate number 4.6 (a & b) showed the stereo image of the fruit callus formed in 2, 4-D (3 mg/l) and 2, 4-D+Kin (1+1 mg/l).

Table 4.5: Effect of auxins singly or in combination with the cytokinins on the growth and nature of fruit callus in *Vinca rosea*.

S. No.	Treatment	Media	Growth hormone (mg/l)	Concentration (mg/l)	Response (%)	Days required for callus initiation	Callus proliferation	Nature of callus
1	T29		Control (Basal)	-	20	20.4 ± 0.55^a	V. Low	White and soft
2	T30	MS	2,4-D	1.5	No result	-	-	-
3	T31			2.0	No result	-	-	-
4	T32			3.0	40	19.8 ± 0.45^b	Low	White and hard
LSD	-	-	-	-	-	0.4740	-	-
5	T33			1+1	60	16.6 ± 0.89^b	Moderate	White and less crystalline
6	T34	MS	2,4-D + Kin	1+2	40	19.4 ± 0.55^a	Low	Creamish Brown and fragile
7	T35			2+1	No result	-	-	-
LSD	-	-	-	-	-	0.8344	-	-

No. of explant inoculated = 10

Each value is the mean of three replicate with standard deviation (mean±S.D). Mean followed by different letters in the same column differ significantly at P=0.05 according to Duncan's new multiple range test.

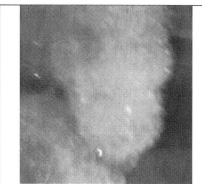 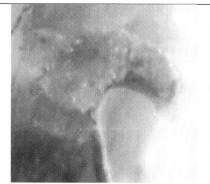

| a: Stereo images of the fruit callus of *V. rosea* in 2, 4-D (3 mg/l). | b: Stereo image of the fruit callus of *V. rosea* on 2, 4-D+Kin (1+1 mg/l). |

Plate No. 4.6 (a & b): **Stereo images of the fruit callus of *V. rosea* on different media.**

4.6: Establishment of cell suspension culture in *Vinca rosea:*

Cell suspension cultures of *Vinca rosea* were tested in MS medium supplemented with one concentration of 2, 4-D (3.0 mg/l) alone and one concentration of 2, 4-D+Kin (1+1 mg/l) in combination. It can be inferred from the table 4.6 that cell suspension response was maximum (70%) in 2, 4-D+Kin (1+1 mg/l) as compared to the suspensions formed in 2, 4-D (3.0 mg/l) which was only 20%. The cells which formed suspension in 2, 4-D+Kin (1+1 mg/l) were embryogenic in nature. The characteristics of *Vinca rosea* suspension cultured cells formed in 2, 4-D+Kin (1+1 mg/l) varies from compact yellowish green callus with rough texture of soft quality to friable pale yellow colour cells with a fine texture.

Plate number 4.7 (a, b & c) showed the establishment of cell suspension cultures in *Vinca rosea.*

Table 4.6: Establishment of cell suspension culture of *Vinca rosea*.

S. No.	Treatment	Media	Growth hormone (mg/l)	Concentration (mg/l)	Response (%)	Days required for suspension	Intensity of cell suspension formation	Characteristics of suspension cultured cells
1	T36	MS	2, 4-D	3.0	20%	16.6±0.89	- +	Yellowish brown callus of hard texture was transformed to Creamish brown soft cells.
2	T37		2, 4-D + Kin	1+1	70%	15.0±1.00	+ ++	Compact yellowish green callus with rough texture was transformed to friable pale yellow colored cells with a fine texture.

-: Calluses gave no suspensions.

+: Calluses gave suspensions consisting elongated cells which did not grow after subculturing; cells died within few weeks.

++: Calluses gave suspensions consisting of both the embryogenic and non-embryogenic cluster of the cells.

Each value is the mean of three replicate with standard deviation (mean±S.D).

43

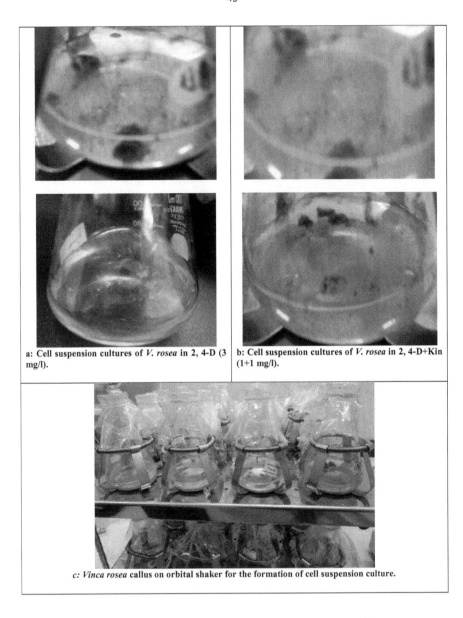

a: Cell suspension cultures of *V. rosea* in 2, 4-D (3 mg/l).

b: Cell suspension cultures of *V. rosea* in 2, 4-D+Kin (1+1 mg/l).

c: Vinca rosea callus on orbital shaker for the formation of cell suspension culture.

Plate No. 4.7 (a, b & c): **Establishment of cell suspension culture of *Vinca rosea*.**

Antimicrobial activity of *Vinca rosea*:

The antibacterial and antifungal activities of *Vinca rosea* were tested against different strains of bacteria and fungi respectively.

Antibacterial activity of *Vinca rosea*:

The antibacterial activity of *in vivo* leaf extract obtained by soxhlet extraction while *in vitro* leaf callus extract, *in vitro* nodal callus extract, *in vitro* fruit callus extract and *in vitro* leaf extract was obtained by simple filtration method, was tested against three different bacterial strains. The methanolic extracts of all the above mentioned parts of plant were prepared. All these extracts were tested against three bacterial strains i.e. *Bacillus subtitles, Bacillus licheniformis and Azotobacter sp.*

The comparison of the antibacterial assay of *in vivo* leaf extract and *in vitro* leaf callus extract of *Vinca rosea* can be seen in the table 4.7. It can be clearly inferred from the table 4.7 that the *in vitro* callus extracts showed better results as compared to the *in vivo* leaf extracts. The maximum zone of inhibition of size 30.3±0.58 mm was shown by *B. subtilis* on the extract concentration of 20 mg/ml where as the zone of inhibition formed by the *in vivo* leaf extract on the same concentration and bacteria was of the size 28.7±1.15 mm. The minimum zone of inhibition of size 2.7±1.15 mm was shown by *Azotobacter sp.* on the extract concentration of 4 mg/ml, on the contrary the *in vivo* leaf callus extract showed no zone of inhibition. The clear difference can viewed from the plate number 4.7.

The data in the table 4.8 revealed the comparison of antibacterial activity of *in vivo* and *in vitro* leaf extracts of *Vinca rosea*. Due to limited *in vitro* leaves, only three extract concentrations were made which were then compared with the *in vivo* leaf extract of those concentrations. Table 4.8 clearly showed that the *in vitro* leaf extracts showed far better results than *in vivo* leaf extracts. The maximum zone of inhibition (30.0±0.00 mm) shown by the *in vitro* leaf extract was at the concentration of 2 mg/ml by the *B. licheniformis* where as the zone of inhibition by *in vivo* leaf extract of the same concentration by the same bacteria was 4.3±0.58 mm. The minimum zone of inhibition (3.0±0.00 mm) was shown by the same bacteria in *in vivo* leaf extract at concentration of 10 mg/ml while at the same concentration *in vitro* leaf extract showed 25.0±0.00 mm zone of inhibition.

The data in the table 4.9 revealed the comparison of antibacterial activity of different extract concentrations of *in vitro* leaf callus, *in vitro* fruit callus and *in vitro* nodal callus of *Vinca rosea*. The *in vitro* leaf callus extract concentrations showed the results as: The maximum zone of inhibition (23.3±1.53 mm) was shown by *B. licheniformis* at the extract concentration of 2 mg/ml. The minimum zone of inhibition of size 3.0±0.00 mm was shown by the same bacteria on the extract concentration of 10 mg/ml. The *in vitro* nodal callus extract showed the results as: the maximum zone of inhibition (28.6±0.57 mm) was formed at the concentration of 2 mg/ml by *Azotobacter sp* whereas the minimum zone of inhibition (11.0±0.00 mm) was formed at the concentration of 10 mg/ml against *B. subtilis*. The *in vitro* fruit callus extract of *Vinca rosea* showed the results as: all the three extract concentrations (2mg/ml, 6mg/ml and 10mg/ml) showed negative results against *Azotobacter sp* while the other two strains showed approximately same results. The maximum zone of inhibition (23.7±1.53) was shown by the extract concentration at 10mg/ml in *B. subtilis* and *B. licheniformis* respectively. The minimum zone of inhibition of size 20.0±0.00 was shown by *B. subtilis* and *B. licheniformis* on the extract concentrations of 2mg/ml and 6mg/ml respectively.

Table 4.7: Comparison of antibacterial activity of different extract concentrations of *in vivo* leaf and *in vitro* leaf callus of *Vinca rosea.*

S. No.	Extract concentration (mg/ml)	Microorganism	*In vivo* leaf zone of inhibition (mm)	*In vitro* leaf callus zone of inhibition (mm)
1	Control	*Bacillus subtilis*	-ive	-ive
		Bacillus licheniformis	-ive	-ive
		Azotobacter sp.	-ive	-ive
2	2	*Bacillus subtilis*	-ive	14.0±1.00
		Bacillus licheniformis	4.3±0.58	23.3±1.53
		Azotobacter sp.	-ive	8.3±1.52
3	4	*Bacillus subtilis*	-ive	26.0±1.73
		Bacillus licheniformis	-ive	26.7±1.53
		Azotobacter sp.	-ive	2.7±1.15
4	6	*Bacillus subtilis*	-ive	20.7±2.08
		Bacillus licheniformis	-ive	11.0±1.73
		Azotobacter sp.	7.3±2.08	11.3±0.58
5	8	*Bacillus subtilis*	-ive	-ive
		Bacillus licheniformis	-ive	17.3±2.08
		Azotobacter sp.	-ive	-ive
6	10	*Bacillus subtilis*	5.7±0.58	16.0±1.73
		Bacillus licheniformis	3.0±0.00	19.0±0.00
		Azotobacter sp.	-ive	-ive
7	12	*Bacillus subtilis*	-ive	20.0±0.00
		Bacillus licheniformis	-ive	10±1.00
		Azotobacter sp.	-ive	-ive
8	14	*Bacillus subtilis*	-ive	17.0±2.65
		Bacillus licheniformis	-ive	-ive
		Azotobacter sp.	13.3±1.53	20.3±0.58
9	16	*Bacillus subtilis*	-ive	-ive
		Bacillus licheniformis	-ive	-ive
		Azotobacter sp.	-ive	11.0±1.00
10	18	*Bacillus subtilis*	-ive	-ive
		Bacillus licheniformis	-ive	-ive
		Azotobacter sp.	9.7±0.58	-ive
11	20	*Bacillus subtilis*	28.7±1.15	30.3±0.58
		Bacillus licheniformis	26.7±1.53	14.3±1.15
		Azotobacter sp.	20.7±0.58	21.3±1.15

Each value is the mean of three replicate with standard deviation (mean±S.D).   -ive= No zone of inhibition (mm).

Table 4.8: Comparison of antibacterial activity of different extract concentrations of *in vivo* leaf and *in vitro* leaf of *Vinca rosea*.

S. No.	Extract concentration (mg/ml)	Microorganism	*In vivo* leaf zone of inhibition (mm)	*In vitro* leaf zone of inhibition (mm)
1	Control	*Bacillus subtilis*	-ive	-ive
		Bacillus licheniformis	-ive	-ive
		Azotobacter sp.	-ive	-ive
2	2	*Bacillus subtilis*	-ive	28.7+1.15
		Bacillus licheniformis	4.3+0.58	30.0+0.00
		Azotobacter sp.	-ive	8.6+0.58
3	6	*Bacillus subtilis*	-ive	17.7+1.15
		Bacillus licheniformis	-ive	26.3+1.52
		Azotobacter sp.	7.3+2.08	12.0+1.00
4	10	*Bacillus subtilis*	5.7+0.58	14.7+1.53
		Bacillus licheniformis	3.0+0.00	25.0+0.00
		Azotobacter sp.	-ive	20.7+0.58

Each value is the mean of three replicate with standard deviation (mean+S.D)

-ive= No zone of inhibition (mm).

Table 4.9: Comparison of antibacterial activity of different extract concentrations of *in vitro* leaf callus, *in vitro* fruit callus and *in vitro* nodal callus of *Vinca rosea*.

S. No.	Extract concentration (mg/ml)	Microorganism	Leaf callus zone of inhibition (mm)	Nodal callus zone of inhibition (mm)	Fruit callus zone of inhibition (mm)
1	Control	*Bacillus subtilis*	-ive	-ive	-ive
		Bacillus licheniformis	-ive	-ive	-ive
		Azotobacter sp.	-ive	-ive	-ive
2	2	*Bacillus subtilis*	14.0+1.00	21.3+0.58	20.0+0.00
		Bacillus licheniformis	23.3+1.53	19.0+1.00	20.7+1.15
		Azotobacter sp.	8.3+1.52	28.6+0.57	-ive
3	6	*Bacillus subtilis*	20.7+2.08	14.0+1.00	21.3+0.58
		Bacillus licheniformis	11.0+1.73	19.3+0.58	20.0+0.00
		Azotobacter sp.	11.3+0.58	25.0+0.00	-ive
4	10	*Bacillus subtilis*	5.7+0.58	11.0+0.00	23.7+1.53
		Bacillus licheniformis	3.0+0.00	17.0+1.73	23.7+1.52
		Azotobacter sp.	-ive	22.3+1.53	-ive

Each value is the mean of three replicate with standard deviation (mean+S.D)

-ive= No zone of inhibition (mm).

49

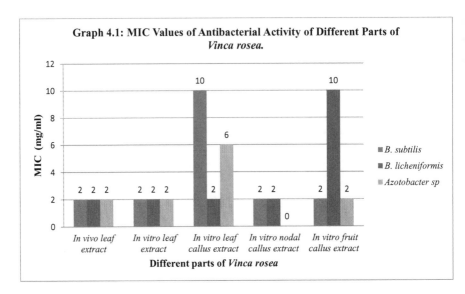

Graph 4.1: MIC Values of Antibacterial Activity of Different Parts of *Vinca rosea.*

Antifungal activity of *Vinca rosea*:

The antifungal activity of different concentrations of *in vivo* leaf extract obtained by soxhlet extraction while *in vitro* leaf callus extract, *in vitro* nodal callus extract, *in vitro* fruit callus extract and *in vitro* leaf extract was obtained by simple filtration method. The methanolic extracts of all the different parts of plant were prepared. All these extracts were tested against three fungal strains i.e. *Aspergillus niger*, *Alternaria solani* and *Rhizopus oryzae*.

The data in the table 4.10 revealed the results of comparison of antifungal activity of different extract concentrations of *in vivo* leaf extract and *in vitro* leaf callus of *Vinca rosea*. The maximum zone of inhibition of size 40.0 ± 1.00 mm was shown by *Aspergillus niger* on the *in vitro* leaf callus extract concentration of 16 mg/ml. On the contrary, 13.0 ± 1.00 mm zone of inhibition was shown by the same fungus at the in *in vivo* leaf extract concentration. The minimum zone of inhibition i.e. 1.00 ± 0.00 was formed by the *in vivo* leaf extract concentration of 2 mg/ml against *Rhizopus oryzae* whereas the minimum zone of inhibition (28.3 ± 1.15) formed by the *in vitro* leaf callus extract was at the concentration of 8 mg/ml against *Aspergillus niger*. No zone of inhibition was formed in the plates containing *Alternaria solani* and showed resistance against all the concentrations of both *in vivo* leaf extracts and *in vitro* leaf callus extracts. *Aspergillus niger* showed varied results as it was resistant to some of the concentrations but not to all. *In vivo* leaf extracts showing zone of inhibition (mm) against *Rhizopus oryzae* can be seen in the plate number 4.8 whereas *in vitro* leaf callus extracts showing zone of inhibitions (mm) against *Aspergillus niger* can be seen in the plate number 4.9.

The data in the table 4.11 revealed the comparison of antifungal activity of different extract concentrations of *in vivo* leaf and *in vitro* leaf of *Vinca rosea*. The two fungal strains i.e. *Aspergillus niger* and *Alternaria solani* showed no zone of inhibitions by the *in vivo* leaf extract concentrations. The *in vivo* leaf extract concentrations i.e. 2 mg/ml, 6 mg/ml and 10 mg/ml showed zone of inhibition 1.0 ± 0.00 mm, 4.3 ± 0.58 mm and 8.0 ± 1.00 respectively against *Rhizopus oryzae*. On the other hand, the two concentrations 2 mg/ml and 6 mg/ml of the *in vitro* leaf callus extract showed no response against all the three fungal strains. At 10 mg/ml, a clear zone of inhibition was formed in all the three fungal strains i.e. *Aspergillus niger* $(5.0\pm1.00$ mm), *Alternaria solani* $(8.0\pm0.00$ mm) and *Rhizopus oryzae* $(10.3\pm0.58$ mm). *In vitro* leaf extract showing zone of inhibition (mm) against different fungi can be seen in the plate number 4.10.

The data in the table 4.12 showed the comparison of antifungal activity of different extract concentrations of *in vitro* leaf callus, *in vitro* fruit callus and *in vitro* nodal callus of *Vinca rosea*. The *in vitro* leaf callus extract showed the results as follows: the two fungal strains *Alternaria solani* and *Rhizopus oryzae* showed complete resistance against all the three concentrations whereas *Aspergillus niger* was not resistant and showed clear zone of inhibitions against all the three concentrations i.e. 2 mg/ml (30.0±0.00 mm), 6 mg/ml (34.6±0.57 mm) and 10 mg/ml (30.3±1.15 mm). The *in vitro* nodal callus extract showed the following results: the maximum zone of inhibition was shown by *Alternaria solani* (18.7±1.15 mm) at 2 mg/l followed by 14.7±0.58 mm and 10.7±0.58 mm at 6 mg/l and 10 mg/l respectively. The minimum zone of inhibition of size 2.0±0.00 mm at 2 mg/l by *R. oryzae* and 2.0±1.00 mm by *A. niger* at 10 mg/l was observed. The *in vitro* fruit callus extract of *Vinca rosea* showed the results as follows: all the three extract concentrations (2 mg/ml, 6 mg/ml and 10 mg/ml) showed negative results against *Rhizopus oryzae* while the other two strains showed approximately same results. The maximum zone of inhibition (8.0±0.00 mm) was shown by the extract concentration at 2 mg/ml in *Alternaria solani* followed by 7.7±0.58 mm in *Aspergillus niger* at the same concentration. The minimum zone of inhibition of size 4.0±1.00 mm was shown by *Alternaria solani* followed by 4.6±0.57 mm in *Aspergillus niger* at 10 mg/ml. *In vitro* nodal callus extracts showing zone of inhibition (mm) against different fungi can be seen in the plate number 4.11 whereas *In vitro* fruit callus extracts showing zone of inhibition (mm) against different fungi can be seen in the plate number 4.12.

Table 4.10: Comparison of antifungal activity of different extract concentrations of *in vivo* leaf and *in vitro* leaf callus of *Vinca rosea*.

S. No.	Extract concentration (mg/ml)	Fungi	In vivo leaf zone of inhibition (mm)	In vitro leaf callus zone of inhibition (mm)
1	Control	Aspergillus niger	-ive	-ive
		Alternaria solani	-ive	-ive
		Rhizopus oryzae	-ive	-ive
2	2	Aspergillus niger	-ive	30.0+0.00
		Alternaria solani	-ive	-ive
		Rhizopus oryzae	1.0+0.00	-ive
3	4	Aspergillus niger	-ive	31.0+1.00
		Alternaria solani	-ive	-ive
		Rhizopus oryzae	4.0+1.00	-ive
4	6	Aspergillus niger	-ive	34.6+0.57
		Alternaria solani	-ive	-ive
		Rhizopus oryzae	4.3+0.58	-ive
5	8	Aspergillus niger	-ive	28.3+1.15
		Alternaria solani	-ive	-ive
		Rhizopus oryzae	6.0+1.73	-ive
6	10	Aspergillus niger	-ive	30.3+1.15
		Alternaria solani	-ive	-ive
		Rhizopus oryzae	8.0+1.00	-ive
7	12	Aspergillus niger	10.7+1.15	37.3+2.30
		Alternaria solani	-ive	-ive
		Rhizopus oryzae	12.3+0.58	-ive
8	14	Aspergillus niger	11.3+1.52	37.0+1.00
		Alternaria solani	-ive	-ive
		Rhizopus oryzae	16.7+1.53	-ive
9	16	Aspergillus niger	13.0+1.00	40.0+1.00
		Alternaria solani	-ive	-ive
		Rhizopus oryzae	20.3+1.15	-ive
10	18	Aspergillus niger	14.0+0.00	29.3+0.57
		Alternaria solani	-ive	-ive
		Rhizopus oryzae	25.0+0.00	-ive
11	20	Aspergillus niger	17.7+0.58	34.7+0.58
		Alternaria solani	-ive	-ive
		Rhizopus oryzae	27.3+2.08	-ive

Each value is the mean of three replicate with standard deviation (mean±S.D)   -ive= No zone of inhibition (mm)

Table 4.11: Comparison of antifungal activity of different extract concentrations of *in vivo* leaf and *in vitro* leaf of *Vinca rosea*.

S. No.	Extract concentration (mg/ml)	Fungi	*In vivo* leaf zone of inhibition (mm)	*In vitro* leaf zone of inhibition (mm)
1	Control	*Aspergillus niger*	-ive	-ive
		Alternaria solani	-ive	-ive
		Rhizopus oryzae	-ive	-ive
2	2	*Aspergillus niger*	-ive	-ive
		Alternaria solani	-ive	-ive
		Rhizopus oryzae	1.0+0.00	-ive
3	6	*Aspergillus niger*	-ive	-ive
		Alternaria solani	-ive	-ive
		Rhizopus oryzae	4.3+0.58	-ive
4	10	*Aspergillus niger*	-ive	5.0+1.00
		Alternaria solani	-ive	8.0+0.00
		Rhizopus oryzae	8.0+1.00	10.3+0.58

Each value is the mean of three replicate with standard deviation (mean+S.D)

-ive= No zone of inhibition (mm)

Table 4.12: Comparison of antifungal activity of different extract concentrations of *in vitro* leaf callus, *in vitro* fruit callus and *in vitro* nodal callus of *Vinca rosea*.

S. No	Extract concentration (mg/ml)	Fungi	Leaf callus zone of inhibition (mm)	Nodal callus zone of inhibition (mm)	Fruit callus zone of inhibition (mm)
1	Control	*Aspergillus niger*	-ive	-ive	-ive
		Alternaria solani	-ive	-ive	-ive
		Rhizopus oryzae	-ive	-ive	-ive
2	2	*Aspergillus niger*	30.0±0.00	9.3±0.58	7.7±0.58
		Alternaria solani	-ive	18.7±1.15	8.0±0.00
		Rhizopus oryzae	-ive	2.0±0.00	-ive
3	6	*Aspergillus niger*	34.6±0.57	4.0±1.73	5.0±0.00
		Alternaria solani	-ive	14.7±0.58	5.0±0.00
		Rhizopus oryzae	-ive	2.1±0.32	-ive
4	10	*Aspergillus niger*	30.3±1.15	2.0±1.00	4.6±0.57
		Alternaria solani	-ive	10.7±0.58	4.0±1.00
		Rhizopus oryzae	-ive	-ive	-ive

Each value is the mean of three replicate with standard deviation (mean±S.D)

-ive= No zone of inhibition (mm)

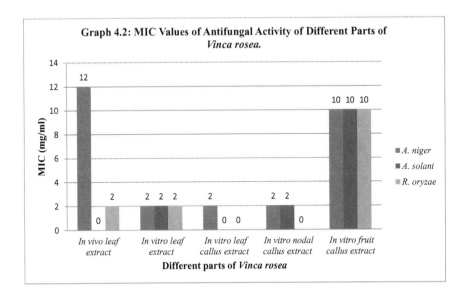

Graph 4.2: MIC Values of Antifungal Activity of Different Parts of *Vinca rosea.*

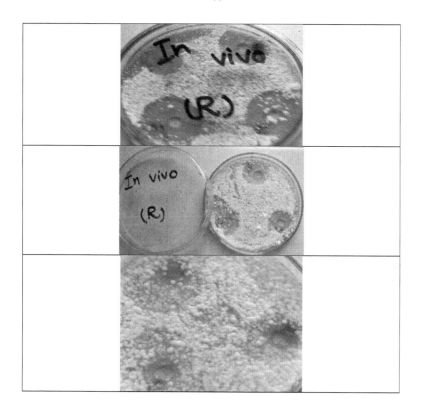

Plate No. 4.8: *In vivo* leaf extracts showing zone of inhibition (mm) against *Rhizopus oryzae*.

57

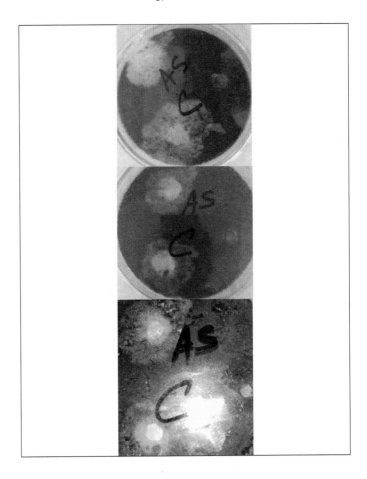

Plate No. 4.9: *In vitro* **leaf callus extracts showing zone of inhibitions (mm) against** *Aspergillus niger.*

58

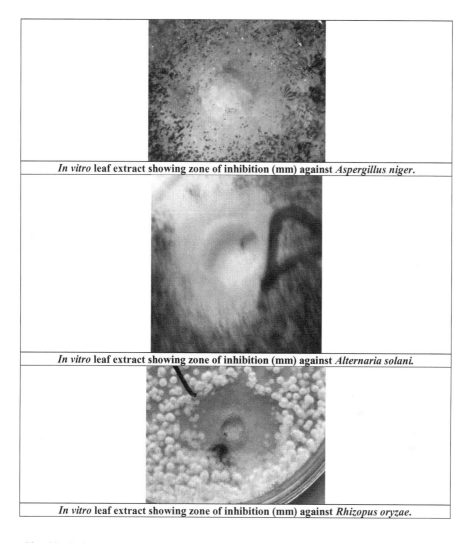

In vitro leaf extract showing zone of inhibition (mm) against *Aspergillus niger.*

In vitro leaf extract showing zone of inhibition (mm) against *Alternaria solani.*

In vitro leaf extract showing zone of inhibition (mm) against *Rhizopus oryzae.*

Plate No. 4.10: *In vitro* leaf extracts showing zone of inhibition (mm) against different fungi.

In vitro nodal callus extracts showing zone of inhibition (mm) against *Aspergillus niger.*

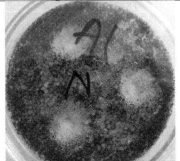

In vitro nodal callus extracts showing zone of inhibition (mm) against *Altrenaria solani.*

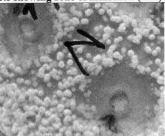

In vitro nodal callus extracts showing zone of inhibition (mm) against *Rhizopus oryzae.*

Plate No. 4.11: *In vitro* **nodal callus extracts showing zone of inhibition (mm) against different fungi.**

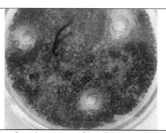

In vitro fruit callus extracts showing zone of inhibition (mm) against *Aspergillus niger*.

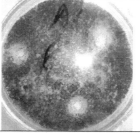

In vitro fruit callus extracts showing zone of inhibition (mm) against *Alternaria solani.*

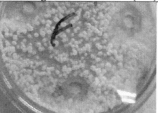

In vitro fruit callus extracts showing zone of inhibition (mm) against *Rhizopus oryzae.*

Plate No. 4.12: *In vitro* fruit callus extracts showing zone of inhibition (mm) against different fungi.

Discussion

The present study is concerned with the micropropagation and callogenesis of medicinally important plant *Vinca rosea*. Various media with different concentrations were tested to optimize the medium showing best growth. The present study also revealed the *in vivo* and *in vitro* comparison of the plant through antimicrobial activity.

Micropropagation of *Vinca rosea*:

The *in vitro* method of producing large number of plants possess great tendency for the production of high quality medicines (Murch *et al.*, 2000). *In vitro* micropropagation is the method by which we can produce novel plants of this desire. The rate of multiplication can be accelerated through micropropagation and it also allows the production of diseased free plants (Khalafallah *et al.*, 2007). The present study is concerned with the establishment of an efficient protocol for *in vitro* micropropagation of *Vinca rosea*.

Various concentrations of cytokinins alone (BAP: 1, 2 & 3 mg/l) or in combination with auxins (BAP+NAA: 1+1, 1+2 & 2+1 mg/l) were used to observe micropropagation. Basal media without growth regulator was also used as control. Table 4.1 and 4.2 showed the effect of MS basal medium, BAP alone and in combination with NAA on shoot propagation from shoot tip and nodal explants of *V. rosea*. BAP (1 mg/l) showed the best results for micropropagation from both shoot tips and nodal portion of *Vinca rosea*. The response was 80-90%. The importance of BAP in multiple shoot formation in *Vinca* and other members of this family are widely reported (Richa *et al.*, 2008; Ravindra *et al.*, 2004). Debergh and Zimmerman (1991) concluded that herbaceous plants are highly responsive to the BAP and this promotes shoot proliferation and produce well formed shoots.

The present study also revealed that among the two explants used, shoot tips induced higher number of shoots per explants as compared to the nodal segment. A similar micropropagation protocol for *Rorippa indica* (a medicinal plant) from shoot tip and nodal explants were reported by Ananthi *et al.*, in 2011.

MS basal medium without any growth regulator did not show any promising results and produced only few shoots per explants. Similar results were observed by Paal *et al.,* (1981) and Cavallini and Lupi (1987).

The combined effect of BAP+NAA was also studied. Table 4.1 and 4.2 showed the combined effect of BAP+NAA (mg/l) on the shoot proliferation of *Vinca rosea* from shoot tips and nodal portion. BAP+NAA (1+1 mg/l & 1+2 mg/l) proved to be the best combinations for both the explants and the response rate was 70%. The combined effect of auxins and cytokinins was also reported by Faheem *et al.,* (2011). They also reported that by increasing or decreasing the amount of BAP beyond the optimal level, the response was delayed and a gradual decrease in the number of shoots was observed.

Callogenesis of *Vinca rosea*:

In the present investigation, different explants were taken to test the different media for the callus induction in *Vinca rosea*. The present study reported the effect of 2, 4-D alone or in combination with Kin on the callus production of *Vinca rosea* from leaf tissues, nodal portion and fruit. Zenk *et al.,* (1977) and Brown (1990) reported that the plant growth hormones such as auxins and cytokinins showed remarkable affects on the growth and differentiation of cultured cells.

Three different concentrations of 2, 4-D was used along with MS basal medium as control. The control did not show any tremendous response and the rate was only 10% to 20% in all the explants. Among the three different concentrations of 2, 4-D (1.5, 2 & 3 mg/l), 3 mg/l showed the best response (60%, 60% & 40%) in all the three explants used. Kalidass *et al.,* (2010) reported that the auxins 2, 4-D supported the growth of callus culture. Morini *et al.,* (2000) reported that the callus production increases with the increasing concentration of 2, 4-D.

The present study also reveals the effect of 2, 4-D+Kin on the callus production. The combination containing 1 mg/l of the each hormone proved to be the best source of callus production in Vinca explants. The response rate was as high as 95%, 80% and 60% in leaf tissues, nodal part and fruit respectively. Taha *et al.,* (2008) reported the production of mass calli from MS medium contained 1 mg/l each of 2, 4-D and Kin. The enhancement in callus initiation and proliferation by using the combinations were also reported by Mollers and Sarkar (1989) by culturing the stem sections from *Catharanthus roseus* on MS medium. Junaid *et al.*, (2006) initiated the embryogenic callus from hypocotyls on the MS medium supplemented with the combination of auxins and cytokinins.

A variety of texture and colour of the callus was produced as mentioned in the present study. The nature of callus also varies with the change in growth regulator. The callus formed was crystalline, less crystalline, fragile, hard, green brown, off-white and white in colour. It was also reported by Abdellatef and Khalafallah (2008). They found that growth regulator type and concentration had a remarkable effect on the callus induction and its physical appearance.

Cell suspension culture of *Vinca rosea:*

The present study also established a cell suspension culture of *V. rosea*. The study revealed that the response of cell suspension formed in 2, 4-D+Kin was 70% as compared with that formed in 2, 4-D (20%) alone. The calli in the former medium gave suspensions consisting elongated cells which did not grow after sub culturing; cells died within few weeks as well as gave suspensions consisting of both the embryogenic and non-embryogenic cluster of the cells whereas the calli in the later medium gave no suspensions or cells that died in few weeks. Kermanee (2004) reported the same results while establishing cell suspension culture of rice. The viability of cell suspension culture is an important system for cell growth studies and secondary metabolite production which was reported by Hilliou *et al.*, (1999) for mass propagation of plants. Nandadeva *et al.*, (1999) reported that embryogenic cell suspension culture is a source and target for gene transformation.

Cells in suspension may exhibit much higher rates of cell division than do cells in callus culture.

To obtain a fine suspension culture, it is important to initiate suspension cultures from a friable callus source (Gurel *et al.*, 2002).

Plant cell cultures are normally established and maintained on media containing auxins and cytokinins. Removal of either hormone from the medium would normally result in culture death (Stafford, 1996).

Antimicrobial activity of *Vinca rosea*:

Medicinal plant products could prove useful in minimizing the adverse effects of various chemotherapeutic agents as well as in prolonging life span and attaining positive general health (Kaushik *et al.*, 2000). It has been evaluated that approximately 75% of the total world population has been using plant-derived medicines (Gaines, 2004). The increasing failure of chemotherapies and antibiotic resistance exhibited by pathogenic microbial infections agents have led to the screening of several medicinal plants for their potential antimicrobial activity (Ritch-Krc *et al.*, 1996; Martins *et al.*, 2001). Antibiotic principles are distributed widely among angiospermic plants. A variety of compounds are accumulated in plant parts accounting for their constitutive antimicrobial activities (Callow, 1983).

As mentioned in the literature earlier, *Vinca rosea* is of high medicinal importance as it possesses the antimicrobial and antifungal activity. It has been suggested that the phytochemical extracts of the plants can be used in allopathic medicine as they are a potential source of antiviral, antitumoral and antimicrobial agents (Nair *et al.*, 2005).

Antibacterial properties of various plant parts, such as leaves, seeds and fruits have been mentioned for some of the medicinal plants for the past two decades (Leven *et al.*, 1979).The effects of plant extracts on the bacterial growth have been studied by many of the researchers in different parts of the world. (Reddy *et al.*, 2001; ErdoÛrul, 2002; Ates and ErdoÛrul, 2003). Table 4.7, 4.8 and 4.9 demonstrated the comparison of antibacterial activity of *Vinca rosea* from different parts i.e. *in vivo* leaf extract versus *in vitro* leaf callus extracts, *in vivo* leaf extract versus *in vitro* leaf extract and *in vitro* leaf callus extract versus *in vitro* nodal callus extract versus *in vitro* fruit callus extract respectively. Tables showed the zone of inhibition (mm) of different extract concentrations against the three bacterial strains i.e. *Bacillus subtilis*, *Bacillus licheniformis* and *Azotobacter sp.* Some strains did not show any zone of inhibition (mm) and the result was negative. Muhammad *et al.*, (2009) reported the antibacterial potential in crude extracts of different parts of *V. rosea* against clinically significant bacterial strains. Marfori *et al.*, (2001) found that the methanol extract from the dual culture of *Trichoderma harzianum* and *Catharanthus roseus* callus showed remarkable antimicrobial activity against the Gram-positive bacteria *Staphylococcus aureus* and *Bacillus subtilis*.

The present study possesses the methanolic extracts. In a study with *V. rosea,* it has been pointed out that the pattern of inhibition largely depends upon extraction procedure, plant part, physiological and morphological state of plant, extraction solvent and microorganism tested. (Goyal *et al.*, 2008). Therefore it can be concluded that the extracts prepared from organic solvents were more active against bacterial species. Similar observations have been reported by Thongson *et al.*, (2004).

Ramya *et al.*, (2008) presented the study that aimed to evaluate the possibility for the presence of novel bioactive compounds against pathogenic bacteria, as most of the pathogens develop drug resistance against commonly used antibiotics. They concluded that Gram negative bacteria were more sensitive when compared to Gram positive bacteria. The study also implicated that bioactive compound of *V. rosea* could potentially be exploited as antibacterial agents. Such results have been shown in the present study too.

Raza *et al.*, (2009) studied the screening of this plant for its antibacterial potential adopting the antibacterial assay. The different parts of *V. rosea* (leaf, stem, flower and root) were used and extracts were subjected to antibacterial assay.

Antifungal activity of *Vinca rosea* has been studied in the present piece of work. Tables 4.10, 4.11 and 4.12 represented the comparison of antifungal activity of *in vivo* leaf extract versus *in vitro* leaf callus extract, *in vivo* leaf extract versus *in vitro* leaf extract and *in vitro* leaf callus extract versus *in vitro* nodal callus extract versus *in vitro* fruit callus extract respectively. In the present study three different fungal strains were used i.e., *Aspergillus niger, Alternaria solani* and *Rhizopus oryzae*. Methanolic extracts of almost all the parts showed clear zones of inhibition (mm). The best response was shown by *Aspergillus niger* followed by *Alternaria solani* and *Rhizopus oryzae*.

Antifungal activity of forty nine botanical extracts were assayed and the data on effect of plant extracts on the growth of *A. niger* was presented by Bobbarala *et al.*, (2009).

Hence it has been supported earlier by Banso and Adeyemo (2007) that the compounds isolated from the medicinal plants possess remarkable toxic activity against bacteria and fungi and possess pharmacological significance.

Jayakumar *et al.*, (2010) concluded that *V. rosea* as an important medicinal plant which might be useful as antioxidant and antimicrobial agents.

Our work also clearly showed that the *in vitro* extract concentrations of all the parts of *V. rosea* exhibit far better results and resistance against both the bacteria and fungi hence possess greater antimicrobial activity *in vitro* as compared with *in vivo* results. Cowan (1999) concluded that plants are rich in secondary metabolites such as tannins, terpenoids, alkaloids, and flavonoids, which have been found *in vitro* to have antimicrobial properties.

The reason behind this may be due to the following:

- *In vitro* plant parts (leaves & calli) were either too young or immature whereas *in vivo* plant parts (leaves) were mature enough.
- *In vitro* part (calli) can be the ideal plant part for the production of antimicrobial compound.
- The alkaloid production may be dependent on undifferentiated and immature cells as compare to the mature and differentiated cells.

The above mentioned reasons have been supported by Roepke *et al.*, (2010) in their paper titled as Vinca drug components accumulate exclusively in leaf exudates of Madagascar periwinkle according to which the entire production of two important alkaloids occurs in young developing leaves. It was also reported that the ability of an alkaloid to inhibit the growth of fungal spores at physiological concentrations found on the leaf surface of *V. rosea* leaves as well as its insect toxicity, provide an additional biological role for its secretion.

Mahroug *et al.*, (2006) also showed that a complex multicellular organization of the MIA biosynthetic pathway occurred in *V. rosea* aerial parts.

St-Pierre *et al.*, (1999) revealed the expression of an alkaloid pathway which follows a basipetal distribution in young expanding leaves. The expression pattern suggested that the alkaloid production was activated in young immature leaves and rapid downfall with the tissue maturation. The accumulation of alkaloids in such tissues may be defensive and protective strategy against predators. (Luijendijk *et al.*, 1996)

The commercial importance of vinblastine and vincristine have led to considerable efforts to produce these chemicals in high yielding callus, cell suspension, and hairy root culture systems, but without apparent success (Van der Heijden *et al.*, 1989). The ability of these culture systems to accumulate alkaloids suggested that the biosynthesis of catharanthine and vindoline is differentially regulated and that vindoline biosynthesis is under a more rigid cell-, tissue-, development- and environment- specific control than that of catharanthine (De Luca, 1993 & Meijer *et al.*, 1993). This hypothesis was further approved by studies that suggested that the ability to synthesize and accumulate vindoline reappeared with the regeneration of shoots and in shoot tissue cultures. (Van der Heijden *et al.*, 1989)

Cowan (1999) concluded the whole study in this way: Ethnopharmacologists, botanists, microbiologists, and natural-products chemists are combing the Earth for phytochemicals and "leads" which could be developed for treatment of infectious diseases.

BIBLIOGRAPHY

- Abdellatef, E. and M. M. Khalafallah. 2008. Influence of growth regulators on callus induction from hypocotyls of medium stapled cotton (*Gossypium hirsutum* L.) cultivar Barac B-67. *J.Soil.Nature*. 2 (1):17-22.

- Ahmed, M. F., S. M. Kazim, S. S. Ghori, S. S. Mahjabeen, S. R. Ahmed, S. M. Ali and M. Ibrahim. 2010. Antidiabetic Activity of *Vinca rosea* Extracts in Alloxan-Induced Diabetic Rats. *International Journal of Endocrinology*, 2010: 1-6.

- Alam, S., N. Akhter, M. F. Begum, M. S. Banu, M. R. Islam, A. N. Chowdhury and M. S. Alam. 2002. Antifungal activities (*in vitro*) of some plant extracts and smoke on four fungal pathogens of different hosts. *Pakistan J. Bio. Sci.*, 5: 307-309.

- Ananthi, P., B. D. Ranjitha Kumari and A. Ramachandran. 2011. *In vitro* propagation of *Rorippa indica* L. from nodal and shoot tip explants. *International Journal for Biotechnology and Molecular Biology Research*, 2(3): 51-55.

- Andrews, J. M. 2001. Determination of minimum inhibitory concentrations. *J. Antimicrob. Chemother.*, 48(1): 5–16.

- Anonymous. Chemical Abstracts. 1988-1993. Columbus: *American Chemical Society*, 44: 119.

- Aslam, J., S. H. Khan, Z. H. Siddiqui, Z. Fatima, M. Maqsood, M. A. Bhat, S. A. Nasim, A. Ilah, I. Z. Ahmad, S. A. Khan, A. Mujib and M. P. Sharma. 2010. *Catharanthus roseus* (L.) G. Don. An important drug: its application and production. *Pharmacie Globale (IJCP).*, 4: 12.

- Antes, D. A. and O. T. ErdoÛrul. 2003. Antimicrobial activities of various medicinal and commercial plant extracts. *Turk J Biol.*, 27: 157-162.

- Azimi, A. A., B. D. Hashemloian, H. Ebrahimzadeh and A. Majd. 2008. High *in vitro* production of ant-canceric indole alkaloids from periwinkle (*Catharanthus roseus*) tissue culture. *African Journal of Biotechnology*, 7 (16): 2834-2839.

- Banso, A. and S. O. Adeyemo. 2007. Evaluation of antibacterial properties of tannins isolated from *Dichrostachys cinerea*. *Afr. J. Biotechnol.*, 6 (15): 1785-1787.

• Batra, J., A. Dutta, M. Jaggi, S. Kumar and J. Sen. 2006. Micropropagation and *in vitro* flowering of medicinal plants, a method for micro breeding. *Floriculture, ornamental and plant biotechnology*, 2: 458-464.

• Bobbarala, V., P. K. Katikala, K. C. Naidu and S. Penumajji. 2009. Antifungal activity of selected plant extracts against phytopathogenic fungi *Aspergillus niger* F2723. *Indian Journal of Science and Technology*, 2(4): 87.

• Brown, T. J. 1990. The Initiation and Mentainance of callus cultures, in Methods in Molecular Biology, Plant cell and Tissue culture (Jeffrey, W.P. and John, M.W., Ed.). *The Human Press*, 6: 57 – 63.

• Callow, J. A. 1983. Biochemical plant phology A Wiley. *Interscience Pub.*

• Cavallini, A. and M. C. Lupi. 1987. Cytological study of callus and regenerated plants of sunflower *(Helianthus annuus L.)*. *Plant Breeding*, 99: 203-208.

• Cowan, M. M. 1999. Plant Products as Antimicrobial Agents. *Clin Microbiol Rev.*, 12(4): 564-582.

• Datta, A. and P. S. Srivastava. 1997. Variation in vinblastine production by *Catharanthus roseus*, during *in vivo* and *in vitro* differentiation. *Pytochemistry*, 46(1): 135-137.

• De Luca, V. 1993. Indole alkaloid biosynthesis: In Methods in Plant Biochemistry. Enzymes of Secondary Metabolism. *Academic Press*, 9: 345–368.

• Debergh, P. C. and R. H. Zimmerman. 1991. Micropropagation Technology and Application. *Kluwer Academic Publishers*, 1-13.

• Debnath, M., C. P. Malik and P. S. Bisen. 2006. Micropropagation: a tool for the production of high quality plant-based medicines. *Curr Pharm Biotechnol.*, 7(1): 33-49.

• El-Sayed, A. and G.A. Cordell. 1981. Catharanthamine: A new antitumor bisindole alkaloid from *Catharanthus roseus*. *J. Nat. Prod.*, 44: 289-293.

• Endress, R. 1994. Plant cell biotechnology. *Springer-Verlag Berlin and Heidelberg GmbH & Co. KG.*, 1st edition.

• ErdoÛrul, O. T. 2002. Antibacterial activities of some plant extracts used in folk medicine. *Pharmaceutical Biol.*, 40: 269-273.

- Faheem, M., S. Singh, B. S. Tanwer, M. Khan and A. Shahzad. 2011. *In vitro* Regeneration of multiplication shoots in *Catharanthus roseus*- An important medicinal plant. *Advances in Applied Science Research*, 2 (1): 208-213.

- Fulzele, D. P. and M. R. Heble. 1994. Large-scale cultivation of *Catharanthus roseus* cells production of ajmalicine in a 20-l airlift bioreactor. *J Biotechnol.*, 35(1): 1–7.

- Gaines, J. L. 2004. Increasing alkaloid production from *Catharanthus roseus* suspensions through methyl jasmonate elicitation. *Pharm. Eng. J.*, 24: 24-35.

- Gamborg, O.L. and G. C. Phillips. 1995. Media preparation and handling. In Plant Cell, Tissue and Organ Culture; Fundamental Methods. *Springer Lab Manual, Springer-Verlag.*, 1–90.

- Ganapathi, B. and F. Kargi. 2011. Recent Advances in Indole Alkaloid Production by *Catharanthus roseus* (Periwinkle). *Journal of Experimental Botany,* 41(3): 259-267.

- Goyal, P., A. Khanna, A. Chauhan, G. Chauhan and P. Kaushik. 2008. *In vitro* evaluation of crude extracts of *Catharanthus roseus* for potential antibacterial activity. *Int. J. Green Pharm.*, 2: 176-181.

- Gurel, S., E. Gurel and Z. Kaya. 2002. Establishment of Cell Suspension Cultures and Plant Regeneration in Sugar Beet (*Beta vulgaris* L.). *Turk. J. Bot.*, 26: 197-205.

- Heijden, V. R., R. Verpoorte, J. G. Hens and H. J. G. Ten-Hoopen. 1989. Cell and tissue cultures of *Catharanthus roseus* (L.) G. Don: a literature survey. *Plant Cell Tissue Organ Cult.*, 18: 231–280.

- Hilliou, P. K., F. P. Christou and M. J. Leech. 1999. Development of an efficient transformation system for *Catharanthus roseus* cell cultures using particle bombardment. *Plant Sci.*, 140: 179-188.

- Hirata. K., A. Yamanaka, N. Kurano, K. Miyamoto and Y. Miura. 1987. Production of indole alkaloids in multiple shoot culture of *Catharanthus roseus* (L.) G. Don. *Agric Biol Chem.*, 51: 1311–1317.

- Jaleel, C. A., R. Gopi, G. M. A. Lakshmanan and R. Panneerselvam. 2006. Triadimefon induced changes in the antioxidant metabolism and ajmalicine production in *Catharanthus roseus* (L.) G. Don. *Plant Science*, 171: 271–276.

- Jayakumar, D., S. J. Mary and R. J. Santhi. 2010. Evaluation of antioxidant potential and antibacterial activity of *Calotropis gigantea* and *Vinca rosea* using *in vitro* model. *Indian Journal of Science and Technology*, 3(7): 720-723.
- Jones. R. N., A. L. Barry and T. L. Gavan. 1985. Washington micro dilution and macro dilution broth procedures. *Manual of Clinical Microbiology*, 972-977.
- Kalidass, C., V. R. Mohan and A. Daniel. 2010. Effect of auxin and cytokinins on vincristine production by callus cultures of *Catharanthus roseus* L. (Apocynaceae). *Tropical and Subtropical Agro ecosystems*, 12: 283 – 288.
- Karuppusamy, S. 2009. A review on trends in production of secondary metabolites from higher plants by *in vitro* tissue, organ and cell cultures. *Journal of Medicinal Plants Research*, 3(13): 1222-1239.
- Kasetsart, J. 2004. Plant Regeneration from Cell Suspension Culture of Rice Varieties Khao Dawk Mali 105 and Suphanburi 1. *Nat. Sci.*, 38: 90 – 96.
- Katzung, B. G. 1995. Basic and Clinical Pharmacology. *Prentice Hall International (UK) Limited*, 6.
- Kaushik P, A. K. Dhiman. 2000. Medicinal Plants and Raw Drugs of India. *Bishen Singh Mahendra Pal Singh*, 12: 623.
- Khalafallah, M. M., E. I. Elgaali and M. M. Ahmed. 2007. *In vitro* multiple shoot regeneration from nodal explants of *Vernonia amygdalina*-An important medicinal plant. *African Crop Science Conference Proceedings*, 8: 747-752.
- Kim, S., N. H. Song, K. H. Jung, S. S. Kwak and J. R. Liu. 1994. High frequency plant regeneration from anther-derived cell suspension cultures via somatic embryogenesis in *Catharanthus roseus*. *Plant Cell Rep.*, 13: 319–322.
- King, P. J. 1984. Induction and maintenance of cell suspension cultures. In: Cell Culture and Somatic Cell Genetics of Plants. *Academic Press*, 1: 130-138.
- Koehn, F. E. and G. T. Carter. 2005. *Nat. Rev. Drug Discover*, 4: 206–220.
- Leven, M., D. A. V. Berghe and F. Mertens. 1979. Medicinal Plants and its importance in antimicrobial activity. *J. Planta Med.*, 36: 311-321.

- Liang, O. P. 2007. Micropropagation and Callus Culture of *Phyllanthus Niruri* L., *Phyllanthus Urinaria* L. And *Phyllanthus Myrtifolius* Moon (Euphorbiaceae) With the Establishment of Cell Suspension Culture of *Phyllantus Niruri* L. *ePrints*, 3-4.

- Luijendijk, T. J. C., E. Vandermeijden and R. Verpoorte. 1996. Involvement of strictosidine as a defensive chemical in *Catharanthus roseus*. *J. Chem. Ecol.*, 22: 1355–1366.

- Marfori, E. C. and A. A. Alejar. 1993. Alkaloid yield variation in callus cultures derived from different plant parts of white and rosy-purple periwinkle, *Catharanthus roseus* (L.) Don. *Philippine J Biotechnol.*, 4: 1–8.

- Marfori, E. C., S. Kajiyama, E. Fukusaki and A. Kobayashi. 2002. Trichosetin, a novel tetramic acid antibiotic produced in dual culture of *Trichoderma harzianum* and *Catharanthus roseus* callus. *Z Naturforsch C.*, 57(5-6): 465-470.

- Martins, A. P., L. Salgueiro, M. J. Goncalves, V. Proencacunha, R. Vila, S. Canigueral and V. Mazzoni. 2001. Essential oil composition and antimicrobial activity of three Zingiberaceae from S. Tomee principle. *J. Planta Med.*, 67: 580-584.

- Meijer, A. M. *et al.* 1993. Regulation of enzymes and genes involved in terpenoid indole alkaloid biosynthesis in *Catharanthus roseus*. *J. Plant Res.*, 3: 145–164.

- Moreno, P. R. H., C. Poulsen, R. Heijden and R. Verpoorte. 1996. Effect of elicitation on different metabolic pathways in *Catharanthus roseus* (L.) G. Don cell suspension cultures. *Enzyme Microb Technol.*, 18: 99–107

- Morini, S., C. D'Onofrio, G. Bellocchi and M. Fisichella. 2001. Effect of 2, 4-D and light quality on callus production and differentiation from *in vitro* cultured quince leaves. *Plant Cell, Tissue and Organ Culture 2000*, 63(1): 47-55.

- Morris, P. 1986. Regulation of product synthesis in cell cultures of *Catharanthus roseus*. III. Alkaloid metabolism in cultured leaf tissue and primary callus. *Planta Med.*, 2: 127–131.

- Muhammad, L. R., N. Muhammad, A. Tanveer and S. N. Baqir. 2009. Antimicrobial activity of different extracts of *Catharanthus roseus*. *Clin. Exp. Med. J.*, 3: 81-85.

- Mukherjee, P. K., P. Balasubramanian, K. Saha, B. P. Saha and M. Pal. 1995. Antibacterial efficiency of *Nelumbo nucifera* (Nymphaeceae) rhizome extract. *Indian Drugs*, 32: 274-276.

- Murashige, T. 1974. Plant propagation through tissue culture. *Ann. Rev. Plant Physiol.*, 25: 135 -166.

- Murashige, T. and F. Skoog. 1962. A revised medium for rapid growth and bioassay with tobacco tissue cultures. *Physiologia Plantarum*, 15: 473-497.

- Murch, S. J., R. S. Krishna and P. K. Saxena. 2000. Tryptophan as a precursor for melatonin and serotonin biosynthesis in *in-vitro* regenerated St. John's ort (*Hypericum perforatum*. cv. Anthos) plants. *Plant Cell Rep.*, 19: 698-704.

- Mustafa, N. R., W. de Winter, F. van Iren and R. Verpoorte. 2011. Initiation, growth and cryopreservation of plant cell suspension cultures. *Nature Protocols,* 6: 715–742.

- Nair, R., T. Kalariya and S. Chanda. 2005. Antibacterial Activity of Some Selected Indian Medicinal Flora. *Turk. J. Biol.,* 29: 41-47.

- Nandadeva, Y. L., C. G. Lupi, C. S. Meyer, P. S. Devi, I. Potrykus and R. Bilang. 1999. Microprojectile-mediated transient and integrative transformation of rice embryogenic suspension cells: effects of osmotic cell conditioning and of the physical configuration of plasmid DNA. *Plant Cell Rep.*, 18: 500- 504.

- Nayak, B. S. and L. M. P. Pereira. 2006. *Catharanthus roseus* flower extract has wound-healing activity in Sprague Dawley rats. *BioMed Central Ltd.,* **6:** 41.

- Nisbet, L. J. and M. Moore. 1997. Will natural products remain an important source of drug research for the future? *Curr Opin Biotechnol.*, 8: 708-712.

- Paal, H. A., E. Kurnik and L. Zabo. 1981. Plantlet regeneration from *in vitro* shoot tip culture of sunflower. *Novenytermeles*, 30: 201-208.

- Paek, K. Y., K. J. Yu, S. I. Park, N. S. Sung and C. H. Park. 1995. Micropropagation of *Rehmannia glutinosa* as medicinal plant by shoot tip and root segment culture. *Acta Horticult.*, 390: 113-20.

- Pahwa, D. 2009. Catharanthus alkaloids. *Scribd, 1-9.*

- Perez, C., M. Pauli and P. Bazerque. 1990. An antibiotic assay by the agar-well diffusion method. *Acta Biol. Med. Exp.*, 2: 708-712.

- Perez-Bermudez, P., H. U. Seitz and I. Gavidia. 2002. A protocol for rapid micropropagation of endangered *Isoplexis*. *In-vitro Cell Dev Biol- Plant*, 38: 178-182.

- Pezzuto, J. 1996. Taxol production in plant cell culture comes of age. *Nature Biotechnol.*, 14: 1083.

- Pietrosiuk, A., M. Furmanowa and B. Lata. 2007. *Catharanthus roseus*: micropropagation and *in vitro* techniques. *Springer Science+Business Media B.V.*, 6: 459–473.

- Pinghin, J. 2003. Your guide to plant cell culture. *The science creative quarterly*, 6.

- Prajakta, J. P. and J. S. Ghosh. 2010. Antimicrobial Activity of *Catharanthus roseus* – A Detailed Study. *British Journal of Pharmacology and Toxicology*, 1(1): 40-44.

- Prakash, S. and J. V. Staden. 2007. Micropropagation of *Hoslundia opposita* Vahl- a valuable medicinal plant. *South African Journal of Botany*, 73: 60-63.

- Ramani, S. and C. Jayabaskaran. *2008*. Enhanced catharanthine and vindoline production in suspension cultures of *Catharanthus roseus* by ultraviolet-B light. *Journal of Molecular Signaling*, 3: 9.

- Ramya, S. *et al.* 2008. *In Vitro* Evaluation of Antibacterial Activity Using Crude Extracts of *Catharanthus roseus* L. (G.) Don. *Ethnobotanical Leaflets*, 2008(1): 140.

- Ravindra K., Kuldeep S., and Veena A. 2004. *In vitro* clonal propagation of *Holarrhena antidysenterica* through nodal explants from mature tree. *In vitro* cellular and developmental biology. Vol. 41 (2): 137-144.

- Raza, M. L., M. Nasir, T. Abbas and B. S. Naqvi. 2009. Antibacterial Activity of Different Extracts from the *Catharanthus roseus*. *Clin. Exp. & Med. J.*, 3(1): 81-85.

- Reddy, P. S., K. Jamil, P. Madhusudhan, *et al.* 2001. Antibacterial activity of isolates from *Piper longum* and *Taxus baccata*. *Pharmaceutical Biol.*, 39: 236-238.

- Richa, B., A. Mohd, A. K. Gaur and P. B. Rao. 2008. *Rauwolfia serpentive* protocol optimization for *in vitro* propagation. *African j. Biotechnology*, 7(23): 4265-4268.

- Ritch-Krc, E. M., N. J. Turner and G. H. Towers. 1996. Carrier herbal medicine an evaluation of the antimicrobial and anticancer activity in some frequently used remedies. *J. Ethnopharmacol.*, 52: 152-156.

75

- Roepkea, J., V. Salima, M. Wua, A. M. K. Thamma, J. Muratab, K. Plossc, W. Bolandc, and V. De Lucaa. 2010. Vinca drug components accumulate exclusively in leaf exudates of Madagascar periwinkle. *PNAS.*, 107(34): 15287–15292.

- Roy, S. K., M. Z. Hossain and M. S. Islam. 1994. Mass propagation of *Rauvolfia serpentina* by *in vitro* shoot tip culture. *Plant Tissue Cult.*, 4: 69-75.

- Sadowska, A. 1991. Plants and natural antitumor compounds. *PWB, Warszawa.* 2-3.

- Shah, M. B. and G. Chauhan. 1996. Recent development of some natural products. In: Supplement to cultivation and utilization of medicinal plants. *SS Honda and MK Koul (eds). RRL.*, 15: 53-96.

- Shariff, N., M. S. Sudarshana, S. Umesha and P. Hariprasad. 2006. Antimicrobial activity of *Rauvolfia tetraphylla* and *Physalis minima* leaf and callus extracts. *African Journal of Biotechnology*, 5 (10): 946-950.

- Shrivastava, R. and P. Singh. 2011. *In vitro* propagation of multipurpose medicinal plant *Gymnema sylvestre* R. Br. (Gudmar). Plant 2002, 38: 178-82.

- Sidhu, Y. 2010. *In vitro* micropropagation of medicinal plants by tissue culture. *The Plymouth Student Scientist*, 4(1): 432-449.

- Stafford, A. 1996. Natural products and metabolites from plants and plant tissue cultures. Plant Cell and Tissue Culture. *Chichester: John Wiley and Sons*, 124-162.

- St-Pierre, B., A. Felipe, F. A. Vazquez-Flota and V. De Luca. 1999. Multicellular Compartmentation of *Catharanthus roseus* Alkaloid Biosynthesis Predicts Intercellular Translocation of a Pathway Intermediate. *The Plant Cell*, 11: 887–900.

- Taha, H. S., M. K. El-Bahr and M. M. S. El-Nasr. 2009. *In vitro* studies on Egyptian *Catharanthus roseus* (L.) G. Don. IV: Manipulation of Some Amino Acids as Precursors for Enhanced of Indole Alkaloids Production in Suspension Cultures. *Australian Journal of Basic and Applied Sciences*, 3(4): 3137-3144.

- Taha, H. S., M. K. El-Bahr and M. M. S. El-Nasr. 2008. *In Vitro* Studies on Egyptian *Catharanthus roseus* (L.) G. Don.: Calli Production, Direct Shootlets Regeneration and Alkaloids Determination. *Journal of Applied Sciences Research*, 4(8): 1017-1022.

- Takeuchi, Y. and A. Komamine. 1982. Effects of culture conditions on cell division and composition of regenerated cell walls in *Vinca rosea* protoplasts. *Plant & Cell Physio.*, 23(2): 249-255.

- Ten-Hoopen, H. J. G., J. L. Vinke, P. R. H. Moreno, R. Verpoorte and J. J. Heijnen. 2002. Influence of temperature on growth and ajmalicine production by *Catharanthus roseus* suspension culture. *Enzyme Microb Technol.*, 30: 56–65

- Ten-Hoopen, H. J. G., W. M. Gulik, J. E. Schlatmann, P. R. H. Moreno, J. L. Vinke, J. J. Heijnen and R. Verpoorte. 1994. Ajmalicine production by cell cultures of *Catharanthus roseus*: from shake flask to bioreactor. *Plant Cell Tissue Organ Cult.*, 38: 85–91.

- Thongson, C., P. M. Davidson, W. Mahakarnchanakul and J. Weiss. 2004. Antimicrobial activity of ultrasound-assisted solvent-extracted spices. *Lett. Appl. Microbiol.*, 39: 401-406.

- Van der Heijden, R. *et al.* 1989. Cell and tissue cultures of *Catharanthus roseus* (L.) G. Don: a literature survey. *Plant Cell, Tissue Organ Cult.*, 18: 231–280.

- Wu. J. and L. Lin. 2002. Elicitor-like effects of low-energy ultrasound on plant (*Panax ginseng*) cells induction of plant defence responses and secondary metabolite production. *Appl. Microbiol. Biotechnol.*, 59: 51-57.

- Yarnell, E. 2004. *Catharanthus roseus* (Madagascar periwinkle): A Natural Antineoplastic and Antidiabetic. A Botanical Quick Review. *Heron botanicals, 1-3.*

- Yuan, Y. J. and Z. D. Hu. 1994. Effect of residual medium on *Catharanthus roseus* callus and suspension cell culture. *Plant Physiol Commun.*, 29: 185–187.

- Zenk, M. H., H. El-Shagi, H. Arens, J. Stockigt, E. W. Weiler, B. Dues. 1977. Formation of indole alkaloids serpentine and ajmalicine in cell suspension cultures of *Catharanthus roseus.* Plant Tissue Culture and its Biotechnological Applications. *Springer*, 27–44.

- Zheng, W. and S. Y. Wang. 2001. Antioxidant Activity and Phenolic Compounds in Selected Herbs. *J. Agric Food Chem.*, 49: 5165-5170.

APPENDICES

APPENDIX 1

Composition of MS (based on Murashige and Skoog, 1962) for *in vitro* micropropagation and callogenesis of *Vinca rosea* L.

Ingredients Macronutrients	MS medium mg/l	Stock concentration mg/l (20X)
NH_4NO_3	16,50	33,000
KNO_3	19,00	38,000
$CaCl_2.2H_2O$	440	8,800
$MgSO_4.8H_2O$	370	7,400
KH_2PO_4	170	3,400
Micronutrients	**MS medium mg/l**	**Stock concentration mg/l (100X)**
$MnSO_4.4H_2O$	22.3	2230
$ZnSO_4.H_2O$	8.6	860
H_3BO_3	6.2	620
$NaMoO_4.2H_2O$	50	5000
$CuSO_4.5H_2O$	0.025	2.5
$CoCl_2.6H2O$	0.025	2.5
KI	0.83	83
Iron Stock	**MS medium mg/l**	**Stock concentration mg/l (200X)**
$FeSO_4.7H2O$	5.56	5,560
Na_2-EDTA	7.40	7,460
Vitamins	**MS medium mg/l**	**Stock concentration mg/l(100X)**
Glycine	2.0	200
Nicotinic acid	0.5	50
Pyridoxine- HCl	0.5	50
Thiaamine-HCl	0.1	10

Sucrose	30 g/l
Myoinositol	0.1 g/l
pH adjusted to	5.5-5.7
Phytagel	1.5 g/l

APPENDIX 2

MS medium formation for Micropropagation and Callogenesis in *Vinca rosea* L.

Constituents	Stock concentration	MS medium
Macronutrients	20X	50 ml/l
Micronutrients	100X	10 ml/l
Vitamins	100X	10 ml/l
Fe-EDTA	200X	5ml/l
BAP		
NAA		
2, 4-D		
Kin		
Sucrose		30 g/l
Myoinositol		0.1 g/l
ppm		1 ml/l
Phytagel		1.5 g/l

pH adjusted to 5.5-5.7

APPENDIX 3

MS media with Phyto Growth Regulators (PGRs) for Micropropagation:

Different concentrations of cytokinins and combinations with auxins used in MS basal medium (Murashige & Skoog, 1962):

Treatment No.	Concentration (mg/l)
T2, T9	MS + 1.0 BAP
T3, T10	MS + 2.0 BAP
T4, T11	MS + 3.0 BAP
T5, T12	MS + 1.0 BAP + 1.0 NAA
T6, T13	MS + 1.0 BAP + 2.0 NAA
T7, T14	MS + 2.0 BAP + 1.0 NAA

APPENDIX 4

MS media with Phyto Growth Regulators (PGRs) for Callogenesis:

Different concentrations of auxins and combinations with cytokinins used in MS basal medium (Murashige & Skoog, 1962):

Treatment No.	Concentration (mg/l)
T16, T23, T30	MS + 1.5 2, 4-D
T17, T24, T31	MS + 2.0 2, 4-D
T18, T25, T32	MS + 3.0 2, 4-D
T19, T26, T33	MS + 1.0 2, 4-D + 1.0 Kin
T20, T27, T34	MS + 1.0 2, 4-D + 2.0 Kin
T21, T28, T35	MS + 2.0 2, 4-D + 1.0 Kin

Printed in Great Britain
by Amazon

48061424R00056